COLLINS GEM
HORSES & PONIES

COLLINS GEM
INSECTS

COLLINS GEM
KINGS & QUEENS

COLLINS GEM
MUSHROOMS & TOADSTOOLS

COLLINS GEM
SNAKES

COLLINS GEM
SPIDERS

COLLINS GEM
STRESS
Survival Guide

COLLINS GEM
TAROT

COLLINS GEM
WINE
Guide

COLLINS GEM
WORLD
atlas

COLLINS GEM
YOGA

COLLINS GEM
ZODIAC
Types

The M24 Chaffee light tank served on after World War II

soon to be followed by many similar establishments. The only problem for the Americans was to decide what to build. The light tank requirement was soon met by the Light Tank series which culminated in the M5. A new heavy tank, ultimately the Heavy Tank M6, was designed from scratch. As for medium tanks, the first attempt was the Medium Tank M2. Designed at a time when little was known of armoured developments elsewhere, the M2 emerged as a mechanically reliable vehicle with plenty of 'stretch' potential but limited by a 37 mm (1.46 in) main turret gun armament and no less than four machine gun sponsons around the upper hull, plus another two machine guns in fixed forward-firing positions on the turret sides.

Multi-turret designs

Multiple armament installations were a feature of tank design during the 1930s. The American M2 was not alone in adopting this 'landship' approach, for one of the most influential of all the between-the-wars designs, the Vickers Independent, featured multiple armament turrets so the M2 was already following a well-trodden path. But multiple weapon stations absorbed manpower; operators often had little else to do other than man a machine gun in cramped and uncomfortable conditions. For the tank commander, directing all these weapons in a cramped turret with limited all-round vision was never easy. Besides, machine guns were of limited value in tank warfare . So the M2 was obsolete before it had entered service - hence the uproar following the reports from the US observers in France. If the Panzer IV had a 75 mm (2.95 in) gun, American tanks would do too. The M2 was abandoned after fewer than 100 had been made .

There was no way the M2 hull could accommodate a turret ring large enough to absorb the recoil. A total hull redesign was needed but that would take time.

The Sherman Firefly carried the 17 pdr anti-tank gun

An interim design was therefore adopted as a way of getting the 75 mm gun into service quickly. This placed the gun in a sponson on the right-hand side of a much-revised M2 hull while retaining the chassis and engine installation of the M2. The result became the Medium Tank M3.

The Fall of France left the British Army with virtually no tanks. When a British purchasing mission saw the new M3 they immediately requested as many vehicles could be provided and even went to the extent of asking for special modifications to suit British requirements. The result was the Grant tank,

23

unmodified M3s delivered to the British were known as the Lee. While the limited traverse of the side-mounted main gun limited the tank's effectiveness, in the Western Desert battles of 1941 and 1942 the Grants and Lees were a valuable asset for the British. So began a reliance on American-supplied armour that was to last throughout the war.

The greatest US tank of the war years was the Medium Tank M4 series, widely known as the Sherman. The M4 was the tank that was produced in greatest numbers in the West, but it was not an all-round success in battle. It was high, lacked sufficient armour and the gun armament was always one step behind the German equivalent. The M4 owed its success to simply being available in huge numbers on a scale the Germans could never hope to match. If the Germans lost a tank it was a major task for them to replace it. If the Allies lost an M4 it was no problem (except to the crew) since it could be replaced without difficulty. The M4 was as much a war-winner as its Soviet counterpart, the T-34 series.

Soviet armoured might

At first the Red Army depended on imported designs such as the Vickers 6-ton tank. Soviet designers were happy to adopt design features from anywhere and took to the American Christie suspension with a will,

A Russian T-34/76 during the battle of Leningrad

developing their BT 'fast tank' series which
culminated in the BT-7. The same suspension was
also utilised for the revolutionary tanks which became
the T-34 series. When the Germans encountered the
T-34/76 during their 1941 invasion of the Soviet
Union they realised that the Panzer IV had met its
match. The T-34/76 was armed with a 76.2 mm (3
in) gun, was well protected, mechanically reliable and
fast. Wide tracks enabled it to cross soft ground
impassable for other tanks. The T-34/76 had another

Factory workers hand over new KV-1s to the Red Army

advantage over the Germans in that it was coming off the production lines in thousands, so much so that by 1945 more T-34 tanks had been produced than any other type, including the American M4. It was just as well: in the intense fighting, a new tank lasted an average of just 7 days at the front. The overall finish of welds and many components was often rough but to the Soviets that mattered little. What did matter was that the T-34/76 was available in quantity. Although the deep German advances of 1941 caused the evacuation of the Soviet tank industry to behind the Urals, tank production continued. Under dreadful conditions, the new factories were built around the machine tools which had been hurriedly shifted East to open field sites. Even as the 'Move to the East' was in progress a new T-34 model was under development. This was the T-34/85, with an 85 mm (3.35 in) gun turret originally designed for a heavy tank, the KV-85. The KV series originally had the same gun and power unit as the T-34 series but gradually evolved into a new family of powerful heavies culminating in the late war years IS-2 and IS-3 with their potent 122 mm (4.80 in) main armament. The KV series were true heavy tanks, intended for breaking through defended lines and destroying enemy tanks. They were formidable vehicles but met their match in the German Tiger.

Tiger and Panther

The only way to counter the T-34 was a new design. In effect the German designers took the best points of the T-34 and added some of their own. The result was the Panther: the best tank design of World War II. The Panther had excellent firepower, protection and mobility. The gun was a long-barrelled high velocity

The Panther tank was designed in response to the T-34

75 mm (2.95 in) piece capable of defeating almost any enemy tank at long ranges: the sloped armour was proof against most Allied anti-armour weapons, and the engine provided as much power as could be desired - the tracks were also wide. However, in their efforts to produce a good tank the German designers produced one which was difficult, expensive, and slow to manufacture.

The Allies came to fear the Panther but it was not invulnerable. The Tiger was a a different matter. This was a design which pre-empted the Panther and was developed during the late 1930s as a breakthrough

A GI stands by the vast bulk of a German Jagdtiger

tank with really thick armour and an 88 mm (3.47 in) main gun. The first Tigers went into action in late 1942 and at first seemed to be as formidable as the designers intended. However, Tigers weighed nearly 60 tonnes, cross-country mobility was poor, and their enormous width meant they could not be loaded onto

29

standard railway wagons without first removing the
tracks and outer road wheels. Tiger production was
slow. But Tigers inflicted damage out of all proportion
to their limited numbers. Post-war analysis indicated
that it took an average of four Allied tank losses to
destroy a single Tiger, but that ratio was offset by the
great efforts needed to get that Tiger into combat; in
addition the four Allied tanks could be readily
replaced - the Tiger could not.

The Tiger II appeared in 1944. Resembling an
enlarged Panther, the Tiger II was armed with an 88
mm (3.47 in) gun and very well armoured. Its
mobility was better than its predecessor but there
were never enough of them. German industry was
already struggling with raw material shortages, a lack
of skilled labour and the effects of Allied bombing.
Just as disruptive was the lack of any centralised
control of tank production. Projects such as the
super-heavy Maus were allowed to continue when
concentration on just a few programmes was what
was needed.

Tank Hunters and Specials

There were two strands in the development of what
became known as tank destroyers (the German term
was Jagdpanzer). One was a lightly armoured, high
mobility chassis with a big gun mounted in a turret.

A Pz IV chassis fitted with an 88 mm anti-tank gun

The other was fit a tank chassis with a thickly armoured box containing a heavier gun than would normally be mounted if the chassis carried a turret. Both types were intended for stalking and destroying enemy tanks, often at extended ranges, rather than tackling them in a stand-up fight.

The American approach was typified by the M10 with its 76.2 mm (3 in) gun in an open-topped turret, later followed by the M36. By contrast the Germans preferred the low silhouette of the Jagdpanther or the little Hetzer, both with 75 mm (2.95 in) guns - the Soviets assumed the same

31

approach with their SU-85 and SU-100 on T-34 chassis.

This use of limited traverse guns in armoured hulls should not be confused with the assault guns such as the German Stug III or Brummbar. They were intended to support infantry operations, even if circumstances often forced them to fight as substitute tanks. Nor should the tank destroyers be confused with self-propelled artillery vehicles such as the Canadian Sexton or the German Hummel. They provided artillery support for armoured formations and were only employed in the direct fire role in dire emergencies since they had only light armour and no overhead protection.

State of the Art, 1945

By 1945 the term tank had come to encompass a wide range of armoured vehicle types. Many armoured chassis which had been intended as tanks ended up carrying out all manner of roles from flame-throwers to bridge-layers. There were also airborne and air defence tanks, supply or personnel carriers, mine-clearing vehicles and mobile observation posts, to name but a few.

By 1945 the earlier German tank supremacy had been reversed by the sheer weight of numbers of the Allied tank formations. From the East the Red Army

IS-II heavy tanks parade through Moscow in 1945

swept forward on waves of T-34s while from the West the M4s, Cromwells and other Allied tanks continued their advances into the Reich. The tanks of 1945 were very different from those of 1939. They were no longer thinly armoured unreliable vehicles of unproven worth but powerful balanced designs. Designs such as the T-34 and Panther had transformed armoured warfare. The tanks of today owe their form to the tanks of 1945.

Sentinel

An Australian Sentinel armed with a 25-pounder gun

By mid 1940 the Japanese threat to Australia was so serious that the Australian government began a crash rearming programme. Since the United Kingdom, then facing a possible German invasion, would be unlikely to supply any tanks, the only course was to develop tanks locally.

Using technical advisers from the United Kingdom it was decided to adopt some commercially available drive components and combine them with locally developed hulls, suspension, chassis and engine arrangements. The result was a remarkable tank known as the AC1, or Sentinel. It featured a cast one-piece hull and cast turret and was powered by a combination of three Cadillac petrol engines driving a common drive shaft. The main armament was limited to a 2-pounder (40 mm) anti-tank gun although there were some later experiments with a 25-pounder (87.6 mm) gun.

The first Sentinel was completed in January 1942, a remarkable engineering achievement considering the limited industrial facilities available. A new factory was built to manufacture the tanks but production ceased in July 1942 after only 66 had been built. By that time it was apparent that the American tank arsenals would be able to meet all future Australian demands so further production would be uneconomic.

Specification

Crew: 5
Engine power: 397 hp (296.16 kW)
Combat weight: 28,450 kg (62,732 lb)
Max speed: 48 km/h (29.8 mph)
Length: 6.325 m (20.75 ft)
Range: 320 km (198.72 miles)
Width: 2.768 m (9.08 ft)
Main gun: 2-pounder (40 mm)
Height: 2.565 m (8.42 ft)
Armour: 25 mm (0.98 in) to 65 mm (2.56 in)

Ram

A Ram medium tank prototype armed with a 2-pounder gun

When Canada entered the war it had only a small army and virtually no tanks. All they were able to obtain were a few American tanks left over from World War I and they were unlikely to get anything from the hard-pressed UK. They therefore decided to produce their own.

The Montreal Locomotive Works created a turreted tank based on components of the American M3 Lee, combined with a Canadian cast steel hull and turret. Superficially, the Ram resembled the American M4 Sherman but its main drawback was its low-powered armament. The turret initially carried a 2-pounder (40 mm) gun, later changed to a 6-pounder (57 mm)

on the Mark 2. By the end of 1941 the first Rams
were entering service and many were shipped over to
Canadian units then stationed in Britain.

However the Ram never saw active service. By the
time it was ready for production the American
arsenals were already supplying sufficient heavier-
armed tanks to meet the changing demands of
armoured warfare. Thus the Rams were diverted to
training or their turrets were removed to convert
them to Ram Kangaroo armoured personnel carriers -
the first of their kind - which saw active service in
Europe during 1944 and 1945.

Other Rams were converted to armoured recovery
vehicles or command posts and an attempt was made
to produce an air defence version, the 'Skink' armed
with four 20 mm cannon. It was not adopted.

Specification - Ram Mk 2

Crew: 5
Engine power: 400 bhp
Combat weight: 29,485 kg (65,014 lb)
Max speed: 40 km/h (24.84 mph)
Length: 5.8 m (19.03 ft)
Range: 230 km (142.83 miles)
Width: 2.9 m (9.51 ft)
Main gun: 2 or 6-pounder (40 or 57 mm)
Height: 2.66 m (8.73 ft)
Armour: 25 mm (0.98 in) to 89 mm (3.51 in)

LT-35

LT-35s were taken over the Germans in 1939

Between the wars the new Czech nation maintained an advanced defence industry with a production capability extending to light tanks. The LT-35, accepted for Czech Army service in 1935 was armed with a Skoda 37 mm (1.46 in) anti-armour gun. It was quite well armed for its size but armoured protection for the three-man crew was no more than adequate. Driving was a difficult process, due in part to the multi-wheel running gear and suspension, resulting in later changes to the steering system.

Unfortunately for the Czech Army they were never

to discover how the LT-35 would prove itself in combat. The 1938 and 1939 take-over of the Czechoslovak state by the Germans resulted in the Czech tanks being incorporated into German Panzer divisions. Despite a Czech decision to phase the LT-35 out of production after 1938, it was extended under German supervision at the CKD works at Skoda and was designated the PzKpfw 35(t). It served in the French campaign of 1940 and the invasion of Russia in 1941. By then the Germans had better tanks in production, so many PzKpfw 35(t)s had their turrets removed and the hulls were converted to create light artillery tractors or mortar and ammunition carriers. A few turreted PzKpfw 35(t)s were handed over to the Bulgarian and Romanian armies.

Specification

Crew: 3
Engine power: 62 hp (46.25 kW)
Combat weight: 8,200 kg (18,081 lb)
Max speed: 30 km/h (18.63 mph)
Length: 4.5 m (14.76 ft)
Range: 110 km (68.31 miles)
Width: 2.08 m (6.82 ft)
Main gun: 37 mm (1.46 in)
Height: 2.3 m (7.55 ft)
Armour: 8 mm (0.32 in) to 25 mm (0.98 in)

LT-38

The LT-38 was a good tank design, soon pressed into German service.

The LT-38 was destined to become one of the most widely used of the Czechoslovak tanks, although not in Czech hands. Ordered into production in 1938, the LT-38 drew on the experiences gained from the earlier LT-35, the main change being the adoption of four large road wheels in place of the LT-35's multi-wheel array. As well as production for the Czech Army, export orders came from Peru, Sweden and Switzerland.

After the German take over, the LT-38 was retained in production until 1942 for the German Army by whom it was known as the PzKpfw 38(t); well over 1400 were built under German supervision. It saw extensive service in France, North Africa, and Russia, but when its utility as a light tank or reconnaissance vehicle was overtaken by the need for heavier armour and a better armament than a Skoda 37 mm (1.46 in) gun and two machine guns the usual German process of adapting the chassis for other purposes commenced. PzKpfw 38(t) chassis were used to carry armaments ranging from 20 mm (0.79 in) Flak cannon to 150 mm (5.91 in) howitzers, as well as 7.5 cm (2.95 in) and 7.62 cm (3 in) anti-tank guns. Other vehicles became light recovery vehicles or supply carriers. More were used for driver training. Perhaps the best of the conversions was the Hetzer tank destroyer.

Specification - PzKpfw 38(t)

Crew: 4
Engine power: 125 hp (93.25 kW)
Combat weight: 9,850 kg (21,719 lb)
Max speed: 42 km/h (26.08 mph)
Length: 4.61 m (15.13 ft)
Range: 250 km (155.25 miles)
Width: 2.14 m (7.02 ft)
Main gun: 37 mm (1.46 in)
Height: 2.25 m (7.38 ft)
Armour: 10 mm (0.39 in) to 25 mm (0.98 in)

FT-17

One of the FT17s manufactured in the USA as the M1917

Introduced in 1917, the Renault FT-17 can lay claim to being the oldest tank in use during World War II. Over 2500 were still in service in 1940, either on second-line Army duties, guarding air bases, or simply stockpiled. As late as 1942 some FT-17s were still deployed by the Vichy French forces in North Africa and Syria.

One reason for the large numbers still around in 1940 was that the FT-17 had been relatively easy to manufacture as it did not have a chassis - the armour plates provided the necessary rigidity. It was a small two-man tank armed with only a 7.5 mm (.30 in) machine gun or a short low-velocity 37 mm (1.46 in) gun. They were also slow, low-powered, had a limited operational range and were thinly armoured. Few carried a radio. (Special radio-equipped FT-17s were supposed to accompany FT-17 formations in action.)

Of the FT-17s captured by the Germans in 1940, many were simply scrapped, their turrets incorporated into the Atlantic Wall. Some were retained for driver training, others for internal security duties in France. Some were deployed during the Fall of Paris in 1944: some had their armament removed and acted as mobile command or observation posts. A very few were fitted with dozer blades to clear snow from Luftwaffe runways.

Specification

Crew: 2
Engine power: 35 hp (26.11 kW)
Combat weight: 7,000 kg (15,435 lb)
Max speed: 7.5 km/h (4.66 mph)
Length: 4.1 m (13.45 ft)
Range: 35 km (21.74 miles)
Width: 1.74 m (5.71 ft)
Main gun: 7.5 mm (.30 in) machine gun or 37 mm (1.46 in) gun
Height: 2.14 m (7.02 ft)
Armour: 6 mm (0.24 in) to 22 mm (0.87 in)

Char B1 bis

Each French armoured division included about 70 Char B1s

The Char B1 bis was a monster, weighing in at 31.5 tonnes (34.5 US tons). It dated from the late 1920s and was intended to be the French Army's main battle tank. The Char B1 bis was considered an advanced vehicle and was easily a match for any of its contemporaries.

The small one-man turret was armed with a 47 mm (1.85 in) gun but the main armament was a front hull-mounted 75 mm (2.95 in) gun, which was aimed using a hydrostatic steering unit operated by the driver since the barrel was fixed for line. Once the gun was aimed and correctly elevated by manual control the driver also fired the gun. Protection was good: in 1940 only the German 8.8 cm (3.47 in) anti-aircraft guns could penetrate its frontal armour.

Production was slow: by 1940 only 400 had been built, partly due to the complexity of the design but mainly because French production was largely by hand - mass production techniques were never introduced. The potential of the Char B1 was wasted as they were committed to battle piecemeal and not concentrated like the German armour. The Germans took a few Char B1 bis into their own service, converting some to take 10.5 cm (4.14 in) howitzers in awkward-looking turrets and installing a hull-mounted flame-thrower in others. Most in German hands were used for driver training.

Specification

Crew: 4
Engine power: 300 hp (223.8 kW)
Combat weight: 31,500 kg (69,457 lb)
Max speed: 28 km/h (17.39 mph)
Length: 6.37 m (20.90 ft)
Range: 150 km (93.15 miles)
Width: 2.5 m (8.20 ft)
Main gun: 1 x 47 mm (1.85 in); 1 x 75 mm (2.95 in)
Height: 2.79 m (9.15 ft)
Armour: 20 mm (0.79 in) to 60 mm (2.36 in)

Captured H 39s used by the Germans in Yugoslavia

The Hotchkiss H 39 was rated as a light (or cavalry) tank and was considered one of the better of the French tanks in 1940. Early models retained the short 1918 pattern 37 mm (1.46 in) gun of the earlier H 35. On later models this was replaced by a longer-barrelled, higher velocity gun, but these had only been issued to platoon commander vehicles by 1940.

Some 400 of the early H 35s were manufactured, followed by over 1,100 of the H 39s. However, while the H 39 was to prove mechanically reliable, like all two-man tanks, it was of limited value in battle, although this was not apparent during the H 39's combat debut in Norway.

After June 1940 the H 39 bagan a second career with the German Army. Fitted with radios, some were included in the Panzer Divisions which invaded Russia in 1941. Thereafter, the H 39 was gradually relegated to second-line and internal security duties while the usual round of conversions was undertaken to produce self-propelled artillery carriers, including 75 mm (2.95 in) anti-tank guns and 105 mm (4.14 in) howitzers, both in awkward-looking open superstructures.

However, enough H 39s in their original form were around to be re-captured in 1945 (some had been retained by the Vichy French) and were among the first tanks to be used by Israel in 1949.

Specification

Crew: 2
Engine power: 120 hp (89.52 kW)
Combat weight: 12,100 kg (26,680 lb)
Max speed: 36.5 km/h (22.67 mph)
Length: 4.22 m (13.85 ft)
Range: 120 km (74.52 miles)
Width: 1.95 m (6.4 ft)
Main gun: 37 mm (1.46 in)
Height: 2.15 m (7.05 ft)
Armour: max 40 mm (1.58 in)

The Renault R35 showing the short 37 mm main gun

The Renault R 35 was supposed to be the replacement for the FT-17 light tank of World War I. The first R 35 prototypes appeared in 1934 and type entered service in 1935. By 1940 some 2,000 had been manufactured, making it numerically the French Army's most important modern tank. It was technically one of the most advanced, having the same horizontal spring suspension design as the AMR 35, a useful top speed and a good overall reliability. Export sales were made to Poland, Turkey, Romania and Yugoslavia.

Unfortunately, the R 35 was let down by two

factors. One was the retention of a short-barrelled 37 mm (1.46 in) main gun dating from 1918, and the second the handicap of only a two-man crew. They were never concentrated to meet the German invasion and many were bypassed, and eventually captured intact. After 1940 they were incorporated into the Panzer Divisions along with a handful of Polish R 35s captured in 1939. A few were handed over to Italy. After 1942 the R 35 became the subject for the usual round of conversions. Turrets were removed to be incorporated into the Atlantic Wall defences, the vehicles serving as artillery tractors. More were converted to carry 10.5 cm (4.14 in) howitzers or 8.1 cm (3.19 in) mortars. About 100 were fitted with 4.7 cm (1.85 in) anti-tank guns to bolster the German anti-invasion defences in France.

Specification

Crew: 2
Engine power: 85 hp (63.41 kW)
Combat weight: 10,600 kg (23,373 lb)
Max speed: 20 km/h (12.42 mph)
Length: 4.02 m (13.19 ft)
Range: 140 km (86.94 miles)
Width: 1.87 m (6.14 ft)
Main gun: 37 mm (1.46 in)
Height: 2.13 m (6.99 ft)
Armour: max 40 mm (1.58 in)

SOMUA S 35

The distinctive SOMUA S 35, one of the best French tanks of 1940

When first revealed in 1935 the SOMUA S 35 was regarded by many as the finest tank in the world. It had a 47 mm (1.85 in) gun in a cast steel turret, a cast steel hull (the first of its kind produced) and it was fast: its 190 hp (141.74 kW) engine imparting a maximum speed of 40 km/h (24.84 mph). By 1940 about 430 S 35s had been manufactured of which 240 were in service. The S35 turret used an electrical drive system which was just as well since the commander had to load, aim and fire the gun.

The one-man turret was to prove a disadvantage in

combat: the commander had to spend much of his time attending to the armament rather than commanding the tank, a fault shared by all the French tanks of 1940. Another weakness exposed in battle was that the cast upper hull was bolted to the lower sections. If an armour-piercing projectile struck the joint between the two, the hull split apart along the length of the vehicle.

After 1940 many S 35s were used by the Germans. Since the cast hull did not lend itself to the usual conversions, they were taken over for second-line duties and driver training; others were passed to Italy. Some were still around in 1944 to counter the Normandy invasion but a few survived to be passed to the Free French who employed them against the Germans once again until the war was over.

Specification

Crew: 3
Engine power: 190 hp (141.74 kW)
Combat weight: 19,500 kg (42,997 lb)
Max speed: 40 km/h (24.84 mph)
Length: 5.38 m (17.65 ft)
Range: 230 km (142.83 miles)
Width: 2.12 m (6.96 ft)
Main gun: 47 mm (1.85 in)
Height: 2.62 m (8.60 fT)
Armour: max 40 mm (1.58 in) hull; 50 mm (1.97 in) turret

Chenillette Lorraine

A captured Chenillette towing a German anti-tank gun

Intended to support tank formations, the Chenillette Lorraine had minimal armour and carried no armament other than the crew's personal weapons. Produced in 1936 it was an advanced concept for that time; little consideration had been given to the resupply of tanks in the field. The were supposed to follow up tank actions and provide ammunition and other supplies, with fuel being carried in small four-wheeled tanker trailers towed behind them. Cargo trailers were used to tow extra supplies.

In 1940 the Chenillette Lorraine could make little contribution to the battle and the Germans captured large numbers of them intact. These were stockpiled for a while before being employed as the basis for many self-propelled guns and other specialist vehicles.

All manner of awkward-looking superstructures were imposed on the Chenillettes to mount guns as large as 15 cm (5.91 in) howitzers for the defence of occupied French territory. Other conversions included 10.5 cm (4.14 in) howitzers, while 4.7 cm (1.85 in) and 7.5 cm (2.96 in) Panzerjagers (tank hunters) were developed. By 1944 a number of vehicles had been manufactured in Vichy France, supposedly for forestry work. After June 1944 these forestry workers were gathered together to act as military supply carriers once again, this time for the Allies. A few served after 1945 and as late as 1956 a handful were still in service with Syria.

Specification

Crew: 2
Engine power: 70 hp (52.22 kW)
Combat weight: 5,740 kg (12,656 lb)
Max speed: 35 km/h (21.74 mph)
Length: 4.19 m (13.75 ft)
Range: 137 km (85.07 miles)
Width: 1.575 m (5.17 ft)
Main gun: none
Height: 1.22 m (4.00 ft)
Armour: basic, 6 mm (0.24 in)

Panzer I

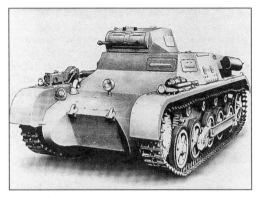

Panzer Is first saw action in the Spanish Civil War

The little Panzer I was supposed to be a training tank for the pre-war Panzer Divisions but it was pressed into use as a light combat tank until at least 1942. The first examples were known as the Landwirtschaftlicher Schlepper (La S), or agricultural tractor, to hide their true purpose until 1938 when the designation Panzer I (or PzKpfw 1) was applied. The first Panzer I was a two-man light tank weighing only 5.5 tonnes (6.06 US tons) and armed with two 7.92 mm (0.31 in) machine guns in a small turret. A second model, the main production version known as the Ausf B, had a more powerful engine and was slightly longer.

The Panzer I first saw action during the German intervention in the Spanish Civil War, 1936-8. By 1939 there were still not enough combat-worthy tanks in German Army service so the Panzer Is were called upon to bulk out the tank units which invaded Poland in 1939 and France in 1940. By 1941 the Panzer I was being withdrawn. Some had already been converted to command vehicles by replacing the turret with a more spacious fixed superstructure while other turretless examples were fitted with a 15 cm (5.91 in) infantry howitzers. In North Africa a few were converted in the field to mount flame-throwers. The most succesful conversion was probably the Panzerjager I, armed with a 47 mm anti-tank gun.

Specification
Crew: 2
Engine power: 100 hp (74.6 kW)
Combat weight: 5,890 kg (12,987 lb)
Max speed: 40 km/h (24.84 mph)
Length: 4.44 m (14.57 ft)
Range: 145 km (90.05 miles)
Width: 2.08 m (6.82 ft)
Main gun: 2 x 7.92 mm (0.31 in) machine guns
Height: 1.73 m (5.66 ft)
Armour: 5 mm (0.20 in) to 35 mm (1.38 in)

Panzer II

A damaged Panzer II being examined by a British officer in the Western Desert

The Panzer II was larger than the Panzer I and was armed with a 2 cm (.79 in) cannon. Intended to be little more than a training tank, it entered frontline service because manufacturing delays held up deliveries of the Panzer III and IV. The first Panzer IIs were issued during 1936, some seeing action in Spain. The first real production model was the Ausf D, followed by the Ausf E, which had an entirely new hull, suspension and armoured layout. Further delays in heavier tank production meant that the last model, the Ausf F, was still on production lines until 1943.

The Panzer II was little more than a reconnaissance vehicle but it was fast, reliable and made up the numbers during the German conquests of 1940-42. Its value against enemy tanks was limited by the poor anti-armour performance of its main armament. After that it was adapted for many other roles. Some, such as the Marder II and Wespe self-propelled guns, were mainly new production vehicles.

Others were conversions, such as the Flamingo flame-thrower with a new and smaller turret for a machine gun - twin flame guns were located on the front hull. Bridgelayer and swimming Panzer IIs were attempted and open-topped 15 cm (5.91 in) howitzer carriers were produced in small numbers. Some Panzer IIs were still around in 1945.

Specification - Ausf F

Crew: 3
Engine power: 140 hp (104.44 kW)
Combat weight: 9,500 kg (20,947 lb)
Max speed: 40 km/h (24.84 mph)
Length: 4.81 m (15.79 ft)
Range: 200 km (124.2 miles)
Width: 2.28 m (7.48 ft)
Main gun: 20 mm (0.79 in) cannon
Height: 2.15 m (7.05 ft)
Armour: 5 mm (0.20 in) to 35 mm (1.38 in)

Luchs

The Luchs was the final development of the Panzer II

The Panzer II Ausf F was not the final development of the Panzer II light tank. Many Panzer II Ausf F models were used in the armoured reconnaissance role for which it was not ideally suited. A new series was developed which looked very different. The first of the new models was the Ausf G which had a set of larger overlapping roadwheels and revised suspension to accommodate them, both increasing the overall height of the vehicle. The larger wheels allowed higher cross-country speeds to be attained but the Ausf F was not produced in significant numbers: it was followed by the Ausf L, H and M, all in small quantities.

Production of the Ausf L began in late 1943. Known as the Luchs (Lynx) it proved to be fast and reliable, with a good cross-country performance. As it was only intended to be a reconnaissance vehicles, there was no need for a heavy armament and it retained the 20 mm (0.79 in) cannon of the Panzer II. Against armoured cars, this was quite sufficient. There was a plan to install a 5 cm (1.97 in) main gun but none were actually delivered to the troops - this version would have been known as Leopard.

Only about 100 Luchs out of 800 ordered were actually manufactured, for production priorities had to be altered as the war entered its final stages. Those that were produced saw the war out on both the Eastern and Western Fronts.

Specification

Crew: 3
Engine power: 180 hp (134.28 kW)
Combat weight: 13,000 kg (28,665 lb)
Max speed: 60 km/h (37.26 mph)
Length: 4.63 m (15.19 ft)
Range: 290 km (180.09 miles)
Width: 2.48 m (8.14 ft)
Main gun: 20 mm (0.79 in) cannon
Height: 2.21 m (7.25 ft)
Armour: 10 mm (0.39 in) to 30 mm (1.18 in)

Wespe

The Wespe carried a 10.5 cm howitzer on a Panzer II chassis

The idea of having self-propelled artillery batteries which could keep pace with tank formations dated back to the 1920s, but it was not until the early 1940s that it was put into practice by the German Army. The withdrawal of the Panzer II from frontline tank units provided the opportunity to use its chassis for self-propelled guns. One of the first projects involved the 7.5 cm (2.96 in) Pak 40 anti-tank gun mounted behind an armoured superstructure on a Panzer II hull. Intended to be a Panzerjager (tank hunter) this arrangement was known as the Marder

II. After about 575 Marder IIs had been produced, production switched to a similar vehicle carrying a 10.5 cm (4.14 in) field howitzer. Produced from early 1943, the Wespe (Wasp) was issued to Panzer division artillery batteries. The hull was lengthened and the suspension modified to absorb the extra firing recoil stresses. The armoured superstructure was open-topped but provided adequate protection for the crew and there was sufficient space to carry 32 rounds. A machine gun was carried for local defence.

About 675 were built before production ceased in July 1943. To support Wespe batteries in the field, 159 unarmed versions were produced as ammunition carriers - the slot in the front shield was plated over although the howitzer could still be mounted if required.

Specification

Crew: 5
Engine power: 140 hp (104.44 kW)
Combat weight: 11,000 kg (24,255 lb)
Max speed: 40 km/h (24.84 mph)
Length: 4.81 m (15.78 ft)
Range: 220 km (136.62 miles)
Width: 2.28 m (7.48 ft)
Main gun: 105 mm (4.14 in) howitzer
Height: 2.3 m (7.55 ft)
Armour: 5 mm (0.20 in) to 30 mm (1.18 in)

Panzer III

A Panzer III Ausf N armed with a short 75 mm gun operating in Russia, 1943

By 1935 the lessons learned with the Panzer 1 and II had provided German industry with the confidence to produce a more ambitious combat tank. Two designs emerged, the first of which became the Panzer III. Armed with a turret-mounted 3.7 cm (1.46 in) gun and two machine guns, it was designed to mount a heavier gun if so required. It had a crew of five; commander, driver, gunner, loader and a radio operator. It was successively up-gunned and up-armoured until at least 20 major variants and numerous sub-marks had been produced. The main armament was increased to 5 cm (1.97 in) and eventually 7.5 cm (2.96 in), using the same mounting as in the early Panzer IV models (q.v.).

Panzer IIIs formed the bulk of the Panzer Divisions' strength during the early war years. The problem was that there were never enough of them. Production was so slow that the Panzer divisions were filled out with Czech tanks in 1940. Manufacture ceased in 1943.

The Panzer III was subjected to the usual round of variants and conversions. The Stug III is the subject of the next entry but there were also command tanks, flame-throwers, 'swimming' tanks meant for the invasion of Britain, artillery observation vehicles and armoured recovery vehicles.

Specification - Ausf F

Crew: 5
Engine power: 300 bhp
Combat weight: 19,800 kg (43,659 lb)
Max speed: 40 km/h (24.84 mph)
Length: 5.38 m (17.65 ft)
Range: 165 km (102.47 miles)
Width: 2.91 m (9.55 ft)
Main gun: 37 mm (1.46 in)
Height: 2.44 m (8.01 ft)
Armour: 12 mm (0.47 in) to 30 mm (1.18 in)

Stug III

A Stug III showing the armoured superstructure and the 75 mm gun and a shield for the roof machine gun

The Sturmgeschutz (assault gun) stemmed from a 1936 requirement for a direct-fire infantry support tank capable of mounting a short-barrelled 7.5 cm (2.96 in) gun on a Panzer III chassis. Entering service in 1940, the Stug III's combat role expanded as German tank production failed to meet demand. Cheaper and quicker to manufacture than turreted tanks, their production became a top priority from 1943 by which time many old Panzer III chassis had been converted to Stug IIIs.

Early models mounted a short, low-velocity gun.

The Ausf F was the first model to mount a higher performance long-barrelled 7.5 cm gun, a variant of the 7.5 cm Pak 40 anti-tank gun. The last production model mounted a 10.5 cm (4.14 in) howitzer - this was the StuH 42. That was not the heaviest armament carried by the Stug III chassis, for one variant, the Stug 33, mounted a 15 cm (5.91 in) howitzer in a raised open superstructure. On all models and variants a roof-mounted machine gun was usually provided for local defence.

Due to their infantry support role the Stug III series were generally better armoured than the Panzer III tank models. Their low silhouette also made them difficult targets. However, on all Stug III models the main armament had only a limited traverse, making them far from ideal for tank combat.

Specification - Ausf G

Crew: 4
Engine power: 300 hp (223.8 kW)
Combat weight: 23,900 kg (52,699 lb)
Max speed: 40 km/h (24.84 mph)
Length: 6.77 m (22.21 ft)
Range: 155 km (96.26 miles)
Width: 2.95 m (9.68 ft)
Main gun: 75 mm (2.96 in)
Height: 2.16 m (7.09 ft)
Armour: 11 mm (0.43 in) to 50 mm (1.97 in)

Panzer IV

An early model Panzer IV with the short 75 mm gun

Although only a few were in service when World War II started, the Panzer IV became the workhorse of the Panzer Divisions. Conceived by General Guderian himself as a heavy support tank for Panzer III units, the Panzer IV was slow to come into production, but when it did appear it soon revealed its fine balance of firepower, mobility and protection.

The first Panzer IVs had a short 7.5 cm (2.96 in) main gun and a five-man crew, all provided with an intercom system, an innovation at that period. Later models, from the Ausf F onwards sported a long 7.5 cm gun. Extra armour was added and innumerable design modifications were introduced over the years - Panzer IV production did not cease until 1945.

It also provided the basis for many other combat vehicles, including the Jagdpanzer IV (next entry), the Nashorn and Hummel self-propelled guns (q.v.) on a lengthened Panzer IV chassis (with 8.8 cm (3.47 in) anti-tank gun and 15 cm (5.91 in) howitzer respectively), and the Brummbar.

The Stug IV was essentially a Stug III superstructure on a Panzer IV chassis. Air defence tanks (Flakpanzer) mounted quadruple 2 cm (0.79 in) or single 3.7 cm (1.46 in) guns in turrets or on open platforms. In addition there were Panzer IV command tanks, submersible tanks, artillery observation vehicles and bridging vehicles.

Specification - Ausf H

Crew: 5
Engine power: 300 bhp
Combat weight: 25,000 kg (55,125 lb)
Max speed: 38 km/h (23.60 mph)
Length: 7.02 m (23.03 ft)
Range: 210 km (130.41 miles)
Width: 2.88 m (9.45 ft)
Main gun: 75 mm (2.96 in)
Height: 2.68 m (8.79 ft)
Armour: 8 mm (0.32 in) to 80 mm (3.15 in)

Jagdpanzer IV

The Jagdpanzer IV's low height made it easier to conceal

Whereas the Stug III was intended to be an infantry support vehicle the Jagdpanzer IV was a tank destroyer. It was a 1943 development consisting of a modified Panzer IV chassis provided with a very thick armoured superstructure, with well sloped platese for added protection, and mounting a limited traverse 7.5 cm (2.96 in) gun in the front plate.

Production of this initial model lasted throughout most of 1944. It proved to be a formidable combat vehicle but the original 7.5 cm gun which was 48 calibres long was replaced by a 7.5 cm L/70 which provided better armour penetration. The extra weight of the longer gun made the chassis nose heavy which wore out the rubber-tyred front road wheels rapidly, and they had to be replaced by all-steel wheels. The result was known as the Panzer IV/70(V).

Yet another Jagdpanzer appeared when the Jagdpanzer IV superstructure and armament was placed directly onto virtually unmodified Panzer IV chassis on production lines - these could be identified by their increased height and were known as Panzer IV/70(A). These, and the Panzer IV/70s, remained in production until the war was in its last stages; over 1,000 were manufactured.

All three of the main Jagdpanzer IV models were issued to the Panzer formations to replace other tank types in short supply. The Jagdpanzer IV was less demanding in production facilities and was quicker to manufacture than a tank with a turret. Despite their considerable firepower, however, they were handicapped by the limited traverse of their gun.

Specification - Jagdpanzer IV

Crew: 4
Engine power: 300 bhp
Combat weight: 25,000 kg (55,125 lb)
Max speed: 40 km/h (24.84 mph)
Length: 6.85 m (22.47 ft)
Range: 210 km (130.41 miles)
Width: 3.17 m (10.40 ft)
Main gun: 75 mm (2.96 in)
Height: 1.85 m (6.07 ft)
Armour: 10 mm (0.39 in) to 80 mm (3.15 in)

Brummbar

The Brummbar was designed to fight in city streets

The Brummbar is one example of the extremes to which the basic Panzer IV chassis could be adapted. With a heavily armoured superstructure housing a short-barrelled 15 cm (5.91 in) howitzer in a ball-mounting on the front plate, it was termed an assault infantry gun. It was developed following the battle of Stalingrad when conventional tanks proved inadequate in the bitter street fighting.

One of the main drawbacks to the early Brummbars, repeated on other similar designs such as the Elefant (q.v.), was that they lacked a defensive weapon against infantry tank-killer squads. A machine gun was carried but this was fired from an open hatch on the superstructure roof, a practice not to be recommended at the combat ranges at which the Brummbar was supposed to operate. Late production models were provided with an extra ball-mounted machine gun in the front superstructure. Most models had provision for stand-off armour plates.

The firepower of the Brummbar at short ranges was considerable but partially limited by the number of rounds which could be carried, probably no more than 30 since the interior was already cramped with the crew of five (four inside the superstructure), their kit, and two radios. About 300 Brummbars were produced, the last of them just before the war ended.

Specification
Crew: 5
Engine power: 300 hp (223.8 kW)
Combat weight: 28,200 kg (62,181 lb)
Max speed: 35 km/h (21.74 mph)
Length: 5.93 m (19.46 ft)
Range: 210 km (130.41 miles)
Width: 2.88 m (9.45 ft)
Main gun: 150 mm (5.91 in)
Height: 2.52 m (8.27 ft)
Armour: 10 mm (0.39 in) to 100 mm (3.94 in)

71

Hummel

A Hummel 15 cm self-propelled howitzer used to support Panzer operations

The Hummel (Bumble-Bee) was a 15 cm (5.91 in) heavy artillery howitzer mounted on a special chassis created by combining features from both the Panzer III and IV chassis. First proposed in July 1942, it shared the same chassis as the Hornisse (Hornet), an extemporised 8.8 cm (3.47 in) anti-tank gun carrier of which 500 were built before the Jagdpanzer IV was available in sufficient numbers.

Both the Hornisse and the Hummel had their ordnance mounted in a large open four-sided superstructure towards the back of a lengthened

Panzer IV hull. The engine was moved forward from its normal position at the rear to a more central location to provide more room at the rear as a fighting compartment.

There was space for only 18 rounds on the Hummel chassis so Hummel batteries were supported by identical ammunition carriers. On the move the Hummel crews travelled in the fighting compartment with only the driver and a radio operator seated under all-round cover at the front. Protection against the weather could be provided by canvas covers. A single machine gun was carried for local defence.

The first Hummels were ready in time for the 1943 Kursk offensive and thereafter each Panzer division was supposed to have at least one Hummel battery.

Specification

Crew: 6
Engine power: 265 hp (197.69 kW)
Combat weight: 24,380 kg (53,757 lb)
Max speed: 42 km/h (26.08 mph)
Length: 7.17 m (23.52 ft)
Range: 215 km (133.51 miles)
Width: 2.97 m (9.75 ft)
Main gun: 15 cm (5.91 in) howitzer
Height: 2.81 m (9.22 ft)
Armour: 10 mm (0.39 in) to 30 mm (1.18 in)

Mobelwagen

A Mobelwagen fully deployed for action

The name Mobelwagen (Furniture Van) was the name given by its crews to a special purpose tank: a Panzer IV chassis converted for the air defence role. The odd name was applied due to the slab-sided appearance of the vehicle when on the move, the 'slabs' being created by the four upward-folding sides which opened horizontally to form a combat platform. Once opened up the main gun could be seen to be an anti-aircraft weapon with a full 360-degree traverse.

Armed with a 3.7 cm (1.46 in) automatic anti-aircraft gun the Mobelwagen was developed to counter growing Allied air supremacy then very apparent on all fronts - 240 Mobelwagens were produced between March 1944 and March 1945.

The Mobelwagen was only one air defence tank based on the Panzer IV chassis. There was also the Wirbelwind (Whirlwind) with four 2 cm (0.79 in) cannon in an open turret and a similar arrangement used for a 3.7 cm gun on the Ostwind (East Wind). The eventual replacement for all these was supposed to be the Kugelblitz (Ball Lightning) with two 3 cm (1.18 in) cannon in an advanced armoured turret but it was developed too late in the war to be produced.

This array of self-propelled anti-aircraft guns proved to be of limited efficiency in the field. They had no centralised fire control equipment - a key element in dealing with low flying enemy aircraft.

Specification

Crew: 6
Engine power: 265 hp (197.69 kW)
Combat weight: 24,380 kg (53,757 lb)
Max speed: 38 km/h (23.60 mph)
Length: 5.92 m (19.42 ft)
Range: 200 km (124.2 miles)
Width: 2.95 m (9.68 ft)
Main gun: 37 mm (1.46 in) anti-aircraft gun
Height: 2.73 m (8.96 ft)
Armour: 10 mm (0.39 in) to 50 mm (1.97 in)

Panther

The Panther strongly influenced post-war western tank designs

The Panther was the best of all the German tanks produced until 1945. It had a remarkable balance of firepower, protection and mobility and the Allies treated Panther formations with great respect, knowing they had a formidable opponent to tackle. The Panther was the Germans' response to the Soviet T-34 (q.v.) which greatly impressed them due to its wide tracks, powerful engine and hard-hitting gun as well as sloping armour which provided extra protection. All these features were built into the Panther which was rushed into production during early 1943.

The Panther was armed with a 7.5 cm (2.96 in) gun 70 calibres long which imparted an excellent anti-armour performance. Overlapping road wheels and a state-of-the-art suspension enabled the Panther to traverse rough terrain at speed. Defensive firepower was balanced with one machine gun in the front hull and a coaxial machine gun in the turret.

The Panther's combat debut at Kursk was a debacle, many breaking down before they got into action, others proving dangerously inflammable. But these problems all stemmed from the hurry to get it into battle; after a few alterations the Panther was developed into a formidable fighting machine with the one disadvantage that there were never enough of them. After World War II Panthers equipped many French Army tank units.

Specification - Ausf G

Crew: 5
Engine power: 690 hp (514.74 kW)
Combat weight: 45,500 kg (100,327 lb)
Max speed: 46 km/h (28.57 mph)
Length: 8.86 m (29.07 ft)
Range: 200 km (124.2 miles)
Width: 3.4 m (11.16 ft)
Main gun: 75 mm (2.96 in)
Height: 2.98 m (9.78 ft)
Armour: 30 mm (1.18 in) to 110 mm (4.33 in)

Jagdpanther

Jagdpanthers knocked out many Allied tanks in Normandy

The Jagdpanther was the tank-hunting equivalent of the Panther, and if the Panther was a formidable opponent the Jagdpanther was even more so. Fast, highly mobile across rough terrain, and difficult to knock out, it was a Panther chassis carrying a well-sloped superstructure mounting a 8.8 cm (3.47 in) Pak 43 anti-tank gun. When firing armour-piercing projectiles, this high-velocity gun could knock out other tanks at combat ranges of well over 1,000 metres, considerably outranging most Allied anti-tank weapons. Combat efficiency was enhanced by a good crew intercom system and convenient racking for the ammunition.

The first of them were ready in June 1944 to

counter the Allied invasion of Normandy but by the time the war ended only just under 400 had been produced, too few to make any difference to the final outcome; production forecasts of 150 a month were never met. Many Jagdpanthers were issued to Panzer battalions in place of tanks.

Nevertheless, the Jagdpanther did achieve considerable local influence on several fronts. Jagdpanthers were instrumental in forcing the initial break-out during the Ardennes offensive of late 1944. But using them as substitute tanks was a mistake. The limited traverse of their main armament was a severe disadvantage if it came to a mobile battle. There were plans to install a 12.8 cm (5.04 in) gun in a Jagdpanther II but the end of the war terminated that project.

Specification

Crew: 5
Engine power: 690 hp (514.74 kW)
Combat weight: 46,000 kg (101,430 lb)
Max speed: 46 km/h (28.57 mph)
Length: 9.9 m (32.48 ft)
Range: 160 km (99.36 miles)
Width: 3.42 m (11.22 ft)
Main gun: 88 mm (3.47 in)
Height: 2.72 m (8.92 ft)
Armour: 25 mm (0.98 in) to 100 mm (3.94 in)

Tiger

A stranded Tiger heavy tank in Normandy, 1944

The massive Tiger dwarfed all previous German tanks. Designed as a heavy assault tank capable of breaking through enemy defences, the project started as early as 1937. After a succession of test rigs, the first recognisable Tiger emerged in early 1942. Its tracks were so wide that a special narrow track was provided for rail transport, prior to which the outer of the overlapping road wheels also had to be removed.

The turret mounted an 8.8 cm (3.47 in) gun 56 calibres long while the frontal armour was 100 mm (3.94 in) thick. The crew was five, commander, gunner, loader, driver and a radio operator who doubled as the gunner for the machine gun in the front hull; another machine gun was mounted coaxially with the main gun.

Tigers first rumbled into action around Leningrad

in August 1942. Soon after, more appeared in Tunisia. Thereafter, they fought on all fronts until the end of the war, a few defending the centre of Berlin in April 1945. 1,354 were manufactured until as late as August 1944.

In action they soon proved to be slow lumbering beasts with a prodigious thirst for fuel, but their heavy armour and powerful gun made them dangerous opponents indeed, especially for Allied tank crew fighting in much less well protected medium tanks like the M4 Sherman. However, while their frontal armour was very difficult to penetrate, they could be destroyed if hit in the rear. Command variants of the Tiger were produced and a few were converted into improvised recovery vehicles.

Specification
Crew: 5
Engine power: 650 hp (484.9 kW) or 700 hp (522.2 kW)
Combat weight: 57,000 kg (125,685 lb)
Max speed: road, 37 km/h (22.98 mph); cross-country, 20 km/h (12.42 mph)
Length: 8.45 m (27.72 ft)
Range: 140 km (86.94 miles)
Width: 3.7 m (12.14 ft)
Main gun: 88 mm (3.47 in)
Height: 2.93 m (9.61 ft)
Armour: 25 mm (0.98 in) to 100 mm (3.94 in)

Sturmmorser Tiger

The Sturmtiger could demolish most buildings with a single shot

The Sturmmorser Tiger, sometimes known as the Sturmtiger, was one of the most unusual armoured vehicles of World War II. It was developed after the battle of Stalingrad during which the German army suffered terrible losses trying to capture heavily-defended buildings. What was needed was a special tank, able to destroy such Russian strongpoints.

The Sturmmorser Tiger was a virtually unmodified Tiger chassis with the front cut away to accommodate a large thickly-armoured box-like superstructure. Mounted in the front plate was a massive RW61 38 cm (14.97 in) breech-loaded mortar which launched a short-range high-explosive rocket projectile. Developed from a naval depth charge launcher, this could demolish virtually any building with a single round. The rocket projectiles weighed 345 kg (760 lb) each and were so large that only 14 could be carried in racks inside the superstructure. Loading the projectiles into the mortar was a slow process while a crane on the back was needed to lift the projectiles into the fighting compartment.

By the time the Sturmmorsers were converted using retired or battle-damaged Tiger tanks (from August to December 1944) the need for such a vehicle/weapon combination was well past. On all fronts, German armies were on the defensive. Only 18 or so conversions were completed.

Specification

Crew: 5
Engine power: 650 hp (484.9 kW)
Combat weight: 65,000 kg (143,325 lb)
Max speed: road, 37 km/h (22.98 mph); cross-country, 20 km/h (12.42 mph)
Length: 6.28 m (20.60 ft)
Range: 140 km (86.94 miles)
Width: 3.57 m (11.71 ft)
Main gun: 380 mm (14.97 in)
Height: 2.85 m (9.35 ft)
Armour: 25 mm (0.98 in) to 150 mm (5.91 in)

Elefant

The giant Elefant tank proved to be a deathtrap for its crews

The Tiger was a Henschel company product, but there was a rival design from Porsche, the Tiger (P). It was rejected due to its complex drive system in which two Panzer IV engines provided power to an electrical transmission for the final drives. The Tiger (P) would have remained at the prototype stage had not Hitler requested an assault tank armed with an 8.8 cm (3.47 in) gun for the Eastern Front. The Porsche design was therefore completed with a heavily armoured superstructure at the rear. It was given several names, including Elefant and Ferdinand. 90 were produced in April and May 1943 and rushed East for the opening of the Kursk offensive that July.

The Elefants went into action on the first day and were a disaster. They had no machine guns for self defence and Russian infantry were able to disable them with anti-tank grenades, mines and petrol bombs. The 8.8 cm gun was awkward to aim. Their drive systems also gave trouble; survivors were returned to the factory and overhauled - they were also provided with a machine gun in the front hull. Five became recovery vehicles. Thereafter the Elefants served mainly in Italy where they proved to be formidable opponents when employed defensively but the local roads could not support their weight and many were eventually destroyed by their own crews.

Specification

Crew: 6
Engine power: 2 x 300 hp (223.8 kW)
Combat weight: 65,000 kg (143,325 lb)
Max speed: 30 km/h (18.63 mph)
Length: 8.14 m (26.70 ft)
Range: 150 km (93.15 miles)
Width: 3.38 m (11.09 ft)
Main gun: 88 mm (3.47 in)
Height: 2.97 m (9.75 ft)
Armour: 30 mm (1.18 in) to 200 mm (7.89 in)

Tiger II

Fortunately for the Allies, fewer than 500 Tiger IIs entered service

The Tiger II was the most powerful combat tank of World War I. It was armed with an 8.8 cm (3.47 in) gun 71 calibres long and was provided with thick armour further improved by sloping plates to give extra protection. Mobility was limited by its great weight and incredible fuel consumption. Nevertheless, the Tiger II was a formidable opponent and Allied troops learned to treat it with the greatest respect, naming it the 'King' or 'Royal' Tiger. Tiger IIs were operated by special Panzer detachments, and the Army and the SS constantly bickered about who should take delivery priority. Wherever the Tiger IIs were used they usually took over the battlefield for they could knock out any Allied tank with ease at considerable ranges. Their own armour was so thick, few enemy weapons could destroy them.

Formidable as the Tiger II was, it suffered from the usual German ailment - there were never enough of them. Production began in early 1944, with the first reaching the Panzer battalions in June that year. But production was constantly disrupted by Allied bombing or delayed by raw material shortages, so that by the time the war ended, only 489 had been manufactured, far too few to stem the Allied advances on all fronts. A few Tiger IIs were produced as command tanks. The Jagdtiger (next entry) was a close variant.

Specification

Crew: 5
Engine power: 600 hp (447.6 kW)
Combat weight: 68,000 kg (149,940 lb)
Max speed: 35 km/h (21.74 mph)
Length: 10.3 m (33.79 ft)
Range: 170 km (105.57 miles)
Width: 3.76 m (12.34 ft)
Main gun: 88 mm (3.47 in)
Height: 3.08 m (10.11 ft)
Armour: 40 mm (1.58 in) to 180 mm (7.09 in)

Jagdtiger

The Jagdtiger: all but invulnerable and armed with a 128 mm gun capable of destroying any Allied tank

The Jagdtiger was based on the Tiger II which already had a powerful 8.8 cm (3.47 in) gun. But the Jagdtiger went one better - it was armed with a 12.8 cm (5.04 in) gun 55 calibre long, by far the most powerful anti-tank weapon used in action during World War II. It could destroy any Allied tank it hit from far beyond the effective range of most Allied guns.

The Jagdtiger was a massive 70 tonne (77.1 US tons) vehicle with its limited-traverse gun mounted in a large fixed turret superstructure - the sides of the turret were sloping plates manufactured in one piece with the sides of the hull. Retaining the Tiger II engine, it was underpowered, but it was to prove a superb defensive weapon platform. Despite the top priority given to Jagdtiger production, it suffered from the same difficulties as the Tiger II - Allied bombing and raw material shortages.

Production began in July 1944, but by the time the War ended only 77 Jagdtigers had been delivered, just sufficient to equip two special Panzer battalions. One battalion was used during the Ardennes offensive and later saw combat in Holland. The other fought in the East, taking part in the desperate defence of Budapest in 1945. Few Jagdtigers survived the war for many were destroyed by their own crews to prevent them being captured intact. One of the few surviving examples is on display at Bovington tank museum.

Specification

Crew: 6
Engine power: 600 hp (447.6 kW)
Combat weight: 70,000 kg (154,350 lb)
Max speed: 37 km/h (22.98 mph)
Length: 10.65 m (34.94 ft)
Range: 170 km (105.57 miles)
Width: 3.63 m (11.91 ft)
Main gun: 128 mm (5.04 in)
Height: 2.95 m (9.68 ft)
Armour: 40 mm (1.58 in) to 250 mm (9.85 in)

Maus

One of the prototypes on evaluation during the war

The Maus was developed by Porsche at the demand of Hitler himself. He demanded a super-heavy tank which would defy anything the Allies could throw at it. What emerged was a monster weighing 188 tonnes (207 US tons)! Its turret was intended to carry not only a 12.8 cm (5.04 in) gun but a 7.5 cm (2.96 in) gun as well. A 15 cm (5.91 in) or 17 cm (6.70 in) gun was planned for later models. Its armour was up to 240 mm (9.46 in) thick, making it virtually invulnerable.

There was a price to pay for all this - when it was first tested in late 1943 the Maus could hardly move.

There was no engine powerful enough to provide the planned-for speed of 20 km/h (12.42 mph). The first prototype had an aircraft engine, supposed to deliver 1200 hp (895.2 kW), but even in ideal conditions the Maus could only creep along at 13 km/h (8.07 mph). Installing a diesel unit in a second prototype made little difference. There was another problem. The Maus was so heavy no bridge could take its weight so it had to be provided with a 'snorkel' arrangement to allow it to wade any rivers in its path. Trials were long delayed by an engine failure on the first prototype. The two prototypes were destroyed in 1945 although another nine were in varying stages of construction. One example was later assembled for exhibition in a museum near Moscow, where it can be seen today.

Specification

Crew: 5
Engine power: 1200 hp (895.2 kW)
Combat weight: 188,000 kg (414,540 lb)
Max speed: 13 km/h (8.07 mph)
Length: 10.09 m (33.11 ft)
Range: 185 km (114.89 miles) (planned)
Width: 3.67 m (12.04 ft)
Main gun: 1 x 128 mm (5.04 in), 1 x 75 mm (2.96 in)
Height: 3.66 m (12.00 ft)
Armour: 40 mm (1.58 in) to 240 mm (9.46 in)

Hetzer

The little Hetzer emerged as one of the best German 'tank hunters'.

Despite its small size and moderate armament, the Jagdpanzer 38, or Hetzer (Baiter) was considered one of the best of the German tank-hunters. Production began in early 1944 at the Skoda works in Pilsen. Using the well-tried suspension and widened chassis of the PzKpfw 38(t), and an up-rated engine, the Hetzer had a low well-sloped upper hull and was armed with a 7.5 cm (2.96 in) anti-tank gun with limited traverse; a machine gun was mounted on the roof in a mounting which could be controlled from within the hull. The interior was somewhat cramped for the four-man crew, but the Hetzer was low, fast and hard-hitting.

Over 2,500 had been produced by 1945. Difficult to spot until it was too late, they were very dangerous opponents for Allied tank crew.

Like most German tanks, the Hetzer was converted into all manner of specialist vehicles. Variants included a flame-thrower tank and an armoured recovery vehicle, both created by re-building Hetzers that had been damaged in action.

Planned self-propelled artillery and air defence versions did not progress beyond paper designs, although one project that did reach the hardware stage featured a 7.5 cm gun with no recoil mechanism. The rigid mounting directly onto the front hull was supposed to absorb all recoil. After the war ended, production continued at Pilsen. The Czech Army adopted the Hetzer and some were later sold to Switzerland. The idea of a small, but well-armed tank destroyer lived on after the war, with new German designs following the Hetzer layout.

Specification

Crew: 4
Engine power: 160 hp (119.36 kW)
Combat weight: 15,750 kg (34,728 lb)
Max speed: 42 km/h (26.08 mph)
Length: 6.38 m (20.93 ft)
Range: 177 km (109.92 miles)
Width: 2.63 m (8.63 ft)
Main gun: 75 mm (2.96 in)
Height: 2.17 m (7.12 ft)
Armour: 8 mm (0.32 in) to 60 mm (2.36 in)

Turan

A Hungarian Turan 1 tank armed with a 75 mm gun

For a relatively small nation, Hungary had an advanced defence industry able to manufacture tanks. One, the Toldi, was a licence-built version of a Swedish Landsverk light tank. Most of the others were light tanks based on Czech Skoda designs, one of which was known as the T-22, later to become the 40M Turan. This design was part of the booty seized by the Germans in their take-over of Czechoslovakia in 1939 and the drawings and tools were simply handed over to the Hungarians for their use.

The 40M Turan was produced in Hungary in two forms. The Turan 1 had a 40 mm (1.58 in) gun created on the same machine tools used to licence-

manufacture 40 mm Bofors Guns in Hungary and
the same ammunition was fired. The Turan 2 had a
75 mm (2.96 in) gun in an enlarged turret. Both
models also carried two 8 mm (0.32 in) machine
guns.

There was also a 40/43M Zrinyi assault gun armed
with a 105 mm (4.14 in) howitzer of local
manufacture on the Turan chassis. Some 300 Turan 1s
were built and 322 Turan 2s, sufficient to equip the
Hungarian Army's only armoured division. Turans
were no match for Soviet armour. From 1943
German tanks, including PzKpfw 38(t)s, Panzer III
and IVs were transferred to the Hungarians instead.
Surviving Turans were overrun in 1944 as Hungary
attempted to break its alliance with Germany.

Specification

Crew: 5
Engine power: 260 hp (193.96 kW)
Combat weight: Turan 1, 18,200 kg (40,131 lb); Turan 2,
18,500 kg (40,792 lb)
Max speed: 47 km/h (29.19 mph)
Length: 5.68 m (18.64 ft)
Range: 165 km (102.47 miles)
Width: 2.54 m (8.33 ft)
Main gun: Turan 1, 40 mm (1.58 in);Turan 2, 75 mm (2.96 in)
Height: 2.33 m (7.64 ft)
Armour: 14 mm (0.55 in) to 50 mm (1.97 in)

CV 33

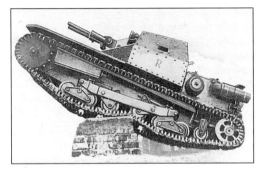

A CV 33 tankette surmounting a low wall obstacle

During the 1920s a series of small two-man tankettes produced by Carden-Loyd attracted a great deal of attention. The Italian Army purchased some and developed their own equivalents. The CV 33 (Carro Veloce 1933) was armed with a single 6.5 mm (0.26 in) machine gun. On the CV 33 Serie, this was changed to two 8 mm (0.32 in) machine guns.

Despite its small size, virtually non-existent protection and limited armament the CV 33 was regarded as a fighting tank by the Italian Army. It evolved into a maid-of-all-work, one of the main variants being a so-called radio (or command) tank - the twin machine guns were retained. There was a flame-thrower tank with the fuel carried in a towed two-wheeled trailer. A more adventurous variant was an assault bridge capable of taking the weight of other tankettes, with the bridge components carried in sections on a trailer behind the CV 33 prior to assembly and laying.

Some machine gun CV 33s were exported to Hungary while others saw action in the Spanish Civil War when Mussolini sent thousands of Italian troops to assist the Nationalists. CV 33s also took part in the invasion of Ethiopia. By the time Italy entered the war in 1940 the CV 33 was already regarded as a death trap yet it was used in the Greek and Albanian campaigns, throughout the early desert fighting and some even took part in the invasion of the Soviet Union in 1941. Thereafter they were withdrawn.

Specification - CV 33 Serie

Crew: 2
Engine power: 43 bhp
Combat weight: 3,435 kg (7,574 lb)
Max speed: 42 km/h (26.08 mph)
Length: 3.16 m (10.37 ft)
Range: 125 km (77.63 miles)
Width: 1.4 m (4.59 ft)
Main gun: 2 x 8 mm (0.32 in) machine guns
Height: 1.28 m (4.20 ft)
Armour: 6 mm (0.24 in) to 13.5 mm (0.53 in)

Carro Armato M11/39

The Italian M11/39 armed with a 37 mm gun in the front hull

The Carro Armato M11/39 was developed as an infantry support tank. The prototype was ready in 1937 and by mid-1940 all 90 of the only batch ordered had been completed. The M11/39 was armed with a 37 mm (1.46 in) gun in a right-hand front superstructure mounting with a limited barrel traverse, with the commander operating a machine gun in a small turret offset to the left. The hull housed the gunner (who had to both aim and load the gun) and the driver.

Some features of the transmission were considered advanced for the period and were to be used on later designs. Another innovation for that period was a Fiat diesel engine. But the suspension was archaic and the small road wheels and narrow tracks hardly made for a smooth ride.

Riveted construction was used throughout and despite the intended infantry support role, armoured protection was modest. By 1940 the M11/39 was obsolete, yet two battalions were shipped to Libya as soon as hostilities commenced. They were used as medium tanks, a role for which they were far from suited, even when opposed by the British cruiser tanks. Against heavy tanks like the Matilda or Valentine, they stood no chance. Many were lost and the Australians used about five captured examples during the fighting around Tobruk in January 1941.

Specification

Crew: 3
Engine power: 105 bhp
Combat weight: 10,970 kg (24,188 lb)
Max speed: 33 km/h (20.49 mph)
Length: 4.74 m (15.55 ft)
Range: 200 km (124.2 miles)
Width: 2.17 m (7.12 ft)
Main gun: 37 mm (1.46 in)
Height: 2.25 m (7.38 ft)
Armour: 7 mm (0.28 in) to 30 mm (1.18 in)

Carro Armato M13/40

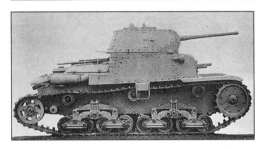

Captured M13/40s were used by British forces in North Africa

By 1940 the drawbacks of the Carro Armato 11/39 (previous entry) were clear and it was decided to enlarge the earlier design and reverse the armament arrangement. The main gun was increased to 47 mm (1.85 in) and moved to an enlarged two-man turret and the machine gun was moved to the hull. Many other automotive aspects of the M11/39 were carried over, including the diesel engine, suspension and road wheels. An initial order was placed in 1940 for 1,900 examples; the final total was 1,960.

The M13/40 proved to be a much more serviceable design especially as the Italian 47 mm anti-tank gun was an excellent weapon. It was accurate and able to penetrate the armour of many British tanks beyond the effective range of their 2 pounder weapons. The first examples were ready for service in North Africa in December 1941. Experience soon revealed the need for greater 'tropicalisation' of engine filters and other components. Later improvements included a more powerful engine, giving rise to the designation M14/41.

The Australians and British often employed captured examples - at one time the British had over 100 in use. Gradually the production of tanks switched to assault guns, the Semovente M40 da 75, with 75 mm (2.96 in) guns of various lengths housed in a low superstructure reminiscent of the German Sturmgeshutz III series (q.v.) and Carro Commando command tanks.

Specification

Crew: 4
Engine power: 105 hp (78.33 kW) (later 145 hp (108.17 kW)
Combat weight: 14,000 kg (30,870 lb)
Max speed: 31.8 km/h (19.75 mph)
Length: 4.915 m (16.13 ft)
Range: 200 km (124.2 miles)
Width: 2.2 m (7.22 ft)
Main gun: 47 mm (1.85 in)
Height: 2.37 m (7.78 ft)
Armour: 14 mm (0.55 in) to 40 mm (1.58 in)

Carro Armato M15/42

The M15/42, last of the Italian World War II tanks

Appearing in late 1942, the Carro Armato M15/42 was supposed to be an improvement on the earlier M13/40 (previous entry). The same general layout was retained but the main armament was an improved 47 mm (1.85 in) high-velocity gun and the engine was changed to a more powerful petrol unit. Armour was thickened, but not enough to match developments elsewhere. Once again, an Italian tank which was already inferior to its likely opponents was ordered into production.

This time the M15/42 production run was short-lived and ceased after only 82 had been made. All the previous basic design drawbacks were carried over, from the modest armoured protection and small calibre main gun to the suspension which rendered the tank slow and uncomfortable. Most of the tanks produced were used in North Africa but nearly all had been lost by the time the battle for Tunisia was over. The same chassis and suspension was utilised to produced the far more practical Semovente M42M da 75/34 assault guns - there was also a Semovente M42L da 105/25 carrying a 105 mm (4.14 in) howitzer. Apart from the Semovente there were command versions of the M15/42 along the same turretless lines as their M13/40 equivalent.

In fact Italian industry abandoned tank manufacture altogether in favour of assault guns and tank destroyers, with the exception of a few experimental designs which were being worked on when Italy was occupied by the German army in 1943.

Specification
Crew: 4
Engine power: 192 hp (143.23 kW)
Combat weight: 15,500 kg (34,177 lb)
Max speed: 40 km/h (24.84 mph)
Length: 5.043 m (16.55 ft)
Range: 220 km (136.62 miles)
Width: 2.23 m (7.32 ft)
Main gun: 47 mm (1.85 in)
Height: 2.385 m (7.83 ft)
Armour: 14 mm (0.55 in) to 50 mm (1.97 in)

103

Type 94

Type 94s took part in the capture of Malaya in 1942

The Japanese managed to produce a very wide array of tanks before and during the war, but most were inferior to Allied tank designs. One of the most widely encountered Japanese tanks of the early war years was the Tankette Type 94 which first appeared in 1932. Originally based on a Vickers commercial design dating from the 1920s, the Type 94 had a two-man crew. The commander in the turret operated the vehicle's only armament, a 7.7 mm (0.30 in) machine gun, but the overall design was so badly thought out that with the gun facing forward the driver could not open his exit hatch. Armour was so thin it could be penetrated by rifle bullets.

The Type 94 was supposed to be a light reconnaissance platform but seems to have spent most of its time acting as an armed tractor towing supplies on a small tracked trailer, although it did see some limited action during the 1930s campaigns in China. One important asset was its narrow width which enabled it to travel along jungle tracks which other vehicles could not negotiate.

After taking part in the advances through Malaya and the early stages of the Burma campaign, the Type 94 was soon withdrawn. Its was obsolete well before Japan entered the war and was retained only because there was little to replace it.

Specification

Crew: 2
Engine power: 32 hp (23.87 kW)
Combat weight: 3,550 kg (7,827 lb)
Max speed: 40 km/h (24.84 mph)
Length: 3.37 m (11.06 ft)
Range: approx 160 km (99.36 miles)
Width: 1.62 m (5.32 ft)
Main gun: 7.7 mm (0.30 in) machine gun
Height: 1.77 m (5.81 ft)
Armour: 4 mm (0.16 in) to 12 mm (0.47 in)

Tankette Type 97

Tankette Type 97 armed here with a 37 mm gun

The Tankette Type 97, or Te-Ke, was the follow-on to the Type 94 (previous entry). Larger, and with a more powerful diesel engine, it had a 37 mm (1.46 in) gun in place of the machine gun - however many Type 97s still retained the machine gun due to a shortage of guns.

Even with these changes the Type 97 was obsolete by the time the first of them were in service. The hull was still so small that there was no space for further crew members, so the commander still had to load and fire the turret gun. But in a gesture towards crew comfort the interior was lined with heat-absorbing asbestos sheets. The suspension, complete with the large trailing wheel, was carried over from the late model Type 94. Armoured protection was still too thin to be proof against small arms fire, and the overall approach remained simple to the point of crudity.

In combat, the Type 97 fared little better than the Type 94. However, since there was little else to hand the Type 97 soldiered on throughout the early war years, distributed in ones and twos in support of infantry operations, but more often used as a forward area supply tractor towing tracked trailers. A few Type 97s were appropriated as artillery observation vehicles. Most had been withdrawn by 1943 or were dug in as pillboxes on Pacific islands.

Specification

Crew: 2
Engine power: 48 hp (35.81 kW)
Combat weight: 4,260 kg (9,393 lb)
Max speed: 45 km/h (27.95 mph)
Length: 3.66 m (12.00 ft)
Range: approx 160 km (99.36 miles)
Width: 1.93 m (6.33 ft)
Main gun: 37 mm (1.46 in)
Height: 1.8 m (5.91 ft)
Armour: 4 mm (0.16 in) to 16 mm (0.63 in)

Light Tank Type 95

The Type 95 was a poor design, but better than most other Japanese tanks

The Light Tank Type 95, also known as the Kyu-Go or Ha-Go, dated from 1935 and owed much to the Tankette Type 94 (q.v.), having the same overall layout and a similar suspension (but with enlarged road wheels and improved springing). However, the crew was increased to three with the extra man acting as hull machine gunner and general mechanic. The commander still served the main turret armament, a 37 mm (1.46 in) gun, and the turret itself was offset to the left to provide more internal space. One of the two 7.7 mm (0.30 in) machine guns carried could fire from a mounting in the turret rear. On some late production models the 37 mm gun was replaced by a 57 mm (2.25 in) low-velocity gun.

By 1939 about 100 had been built. More Type 95s were produced than any other Japanese tank but once in service they were used in penny packets, as infantry support tanks - a role for which they were not suited. Apart from the poor armoured protection, crew visibility was very limited and there was a distinct area around each tank which could not be observed by the crew. The Type 95 hardly had an auspicious career and despite being the most important Japanese tank of World War II, it must be judged a failure. Production ceased in 1943.

Specification

Crew: 3
Engine power: 110 hp (82.06 kW)
Combat weight: 7,620 kg (16,802 lb)
Max speed: 45 km/h (27.95 mph)
Length: 4.38 m (14.37 ft)
Range: approx 160 km (99.36 miles)
Width: 2.05 m (6.73 ft)
Main gun: 37 mm (1.46 in)
Height: 2.18 m (7.15 ft)
Armour: 6 mm (0.24 in) to 12 mm (0.47 in)

Medium Tank Type 89

The Japanese Type 89 was based on a British design

The Medium Tank Type 89 derived from a Vickers commercial design, the Medium Type C. The Japanese Army adopted the Type C in 1929, as a short cut to getting something better than tankettes into service quickly. Production was by Mitsubishi. There were two models which differed appreciably. The initial Type 89A was powered by a petrol unit adapted from an aircraft engine. The Type 89B had a diesel and numerous alterations to the front hull armour plating, plus a revised turret with a cupola for the commander. The track was also changed on the

Type 89B to overcome the excessive track wear of the Type 89A.

The Type 89 was armed with a low-velocity 57 mm (2.25 in) gun with limited anti-armour performance; the turret also had the usual Japanese feature of a 6.5 mm (0.26 in) machine gun firing from a mounting at the back. Another 6.5 mm machine gun was mounted in the front hull.

Despite being larger than most previous Japanese tanks the Tank 89 had a very cramped interior and crew visibility was poor. Throughout its service life the Type 89 series underwent numerous detail changes. A more powerful diesel engine was fitted to late production models. Cross-country mobility was increased by the addition of an unditching tail at the rear.

Specification - Type 89B

Crew: 4
Engine power: 120 hp (89.52 kW)
Combat weight: 12,700 kg (28,003 lb)
Max speed: 24 km/h (14.90 mph)
Length: 5.867 m (19.25 ft)
Range: 160 km (99.36 miles)
Width: 2.16 m (7.09 ft)
Main gun: 57 mm (2.25 in)
Height: 2.59 m (8.50 ft)
Armour: 10 mm (0.39 in) to 17 mm (0.67 in)

111

Medium Tank Type 97

Medium Tank Type 97, the best of all the World War II Japanese tanks

Introduced in 1937, the Medium Type 97, or Chi-Ha, was to prove itself the best of the World War II Japanese tanks, but even so it was no match for any Allied tank. The overall design drew heavily on features of many European tanks, but little attention was paid to the needs of mass production, so output was slow and could never meet demand. It was armed with a 57 mm (2.25 in) gun in an odd-shaped two-man turret, offset to the left of the riveted hull; the machine gun mounting at the back of the turret was retained. The suspension was a considerable improvement over previous Japanese types but still lacked shock absorbers and thus provided a rough ride. Power was provided by a V-12 diesel engine. Armoured protection was poor.

Used on all Japanese fronts during the early part of the war, the Type 97 underwent numerous changes, including a larger turret mounting a 47 mm (1.85 in) high-velocity anti-tank gun from 1942 onwards: this was known as the Shinhoto Chi-Ha. Production of this latter model continued until the war ended. One variant was the Shi-Ki command tank with a 37 mm (1.46 in) gun in the front hull in place of the machine gun; the turret gun was a dummy. The Se-Ri was an ARV mounting a collapsible crane. Other variants included a mine flail carrier, a dozer blade tank which retained its main armament, and various self-propelled artillery equipments.

Specification

Crew: 4
Engine power: 150 hp (111.9 kW) or 170 hp (126.82 kW)
Combat weight: 13,465 kg (29,690 lb)
Max speed: 40 km/h (24.84 mph)
Length: 5.53 m (18.14 ft)
Range: 240 km (149.04 miles)
Width: 2.31 m (7.58 ft)
Main gun: 47mm (1.85 in) or 57 mm (2.25 in)
Height: 2.36 m (7.74 ft)
Armour: 8 mm (0.32 in) to 33 mm (1.30 in)

TK Tankette Series

The TK-3 was hopelessly outclassed by German tanks

The TK series was a Polish development of the Vickers Carden-Loyd Mark IV machine gun carrier design. This model had attracted the attention of the Polish army which bought one and promptly set about developing their own version. The main production version was the TK-3, over 300 of which were built from 1931. A small two-man vehicle powered by a Ford engine, the TK-3 was the first all-Polish tracked armoured vehicle. The crew was housed in a lightly armoured superstructure with a 7.92 mm (0.31 in) machine gun provided for the commander. Later developments included the TKS which was powered by a licence-produced FIAT engine (some 390 produced) and the TKS-D which had a Bofors 37 mm (1.46 in) anti-tank gun mounted on a reinforced front plate - it was not adopted. By 1939 a small number of TK-3s had been modified to mount a 20 mm (0.79 in) cannon in place of the machine gun.

One oddity of the TK-3 was its road transport trailer. The intention was that the TK-3 would mount the trailer via a ramp and then the tankette's final drive would be connected to the trailer wheels, allowing it to travel long distances. It is not known how successful this was, for most long journeys seem to have been covered by carrying TK-3s on flat-bed trucks. In September 1939 the TK series tankettes were hopelessly outclassed by the German Panzers and were simply swept aside.

Specification - TK-3

Crew: 2
Engine power: 40 hp (29.84 kW)
Combat weight: 2,430 kg (5,358 lb)
Max speed: 46 km/h (28.57 mph)
Length: 2.58 m (8.46 ft)
Range: not known
Width: 1.78 m (5.84 ft)
Main gun: 7.92 mm (0.31 in) machine gun
Height: 1.32 m (4.33 ft)
Armour: 4 mm (0.16 in) to 8 mm (0.32 in)

Light Tank 7TP

Polish 7TP tanks take part in the 1939 mobilisation

Although the Light Tank 7TP was initially purchased from Vickers-Armstrong it was later licence-produced in Poland and may thus be regarded as a Polish tank. The Vickers product was the Mark E of their commercial 6-ton tank for which the Poles were one of the main customers - other Vickers 6-ton tanks were sold to Bulgaria, Finland, the Soviet Union (T-26 - q.v.) and Greece, as well as South America and Thailand.

The 7TP was powered by a Polish engine (actually a licence-produced Sauer unit from Switzerland) but was otherwise basically identical to the Vickers original. It was produced in Poland from 1934 onwards. One, the first model, had two small turrets with a 7.92 mm (0.31 in) Browning machine gun in each. The other, introduced in 1937, had a 37 mm (1.46 in) Bofors anti-tank gun in a single turret.

In 1939 two Polish tank battalions were equipped with the 7TP. Although it looked rather dated by 1939 the 7TP was a match for German tanks such as the Panzer I and II. After the surrender, the German Army used as many 7TPs as they could, originally as gun tanks for internal policing in occupied territories but later as artillery tractors.

Specification
Crew: 3
Engine power: 110 hp (82.06 kW)
Combat weight: 9,400 kg (20,727 lb)
Max speed: 37 km/h (22.98 mph)
Length: 4.56 m (14.96 ft)
Range: approx 150 km (93.15 miles)
Width: 2.3 m (7.55 ft)
Main gun: 37 mm (1.46 in) or 2 x 7.92 mm (0.31 in) machine guns
Height: 2.016 m (6.61 ft)
Armour: 5 mm (0.20 in) to 17 mm (0.67 in)

117

T-38 Light Amphibious Tank

T-38s took part in the Russian invasion of Finland

The T-38 light amphibious tank was based on the earlier T-37 which appeared during the early 1930s. The T-37 was based on a Vickers Carden-Loyd design with a strengthened suspension. Power was provided by a Ford-based GAZ AA engine.

The T-37 proved generally satisfactory, but changes were considered necessary and in 1936 work began on the T-38. The T-38 was wider, but lower than the T-37 and had a different transmission. Both the T-37 and T-38 were two-man vehicles with their armament limited to a single 7.62 mm (0.30 in) machine gun. Late production T-38s had a 20 mm (0.79 in) cannon. Their amphibious role meant that weight had to be kept to a minimum and armour protection was therefore poor, making the vehicle a deathtrap in battle.

Over 1,500 T-38s were manufactured and were used in the 1940 Winter War against Finland. The T-38 did gain the distinction of being the first Soviet airborne tank in 1936, slung under a heavy bomber fuselage, but was not used operationally in this role. Few survived the German onslaught in 1941 and the handful captured were handed over to the Luftwaffe for air base perimeter defence. The only known variant of the T-38 was a command vehicle with extra radios - many standard T-38 lacked radios and had to rely on hand signals or flags to transmit reconnaissance reports.

Specification - T-38

Crew: 2
Combat weight: 3,300 kg (7,276 lb)
Length: 3.78 m (12.40 ft)
Width: 3.33 m (10.93 ft)
Height: 1.63 m (5.35 ft)
Engine power: 40 hp (29.84 kW)
Max speed: 40 km/h (24.84 mph)
Range: 170 km (105.57 miles)
Main gun: 7.62 mm (0.30 in) machine gun
Armour: 3 mm (0.12 in) to 9 mm (0.35 in)

T-40 Light Amphibious Tank

Buoyancy tanks made the T-40's hull very bulky

When the shortcomings of the T-37/T-38 light tanks
were recognised, attempts were made to produce
more viable successors. Production was assigned to
factories normally producing automobiles. The two-
man T-40 was a simple design and made extensive use
of automobile components wherever possible,
including the engine and transmission. Early T-40s
were armed with a 12.7 mm (0.50 in) heavy machine
gun, later models had a 20 mm (0.79 in) cannon.

The most noticeable feature of the T-40 was its
bulky boat-shaped hull produced by two flotation
tanks which kept the vehicle afloat in the water, while
a propeller at the rear provided propulsion. The

engine was on the right-hand side of the welded hull.

The first T-40s were used during the latter stages of the Winter War with Finland. Their thin armour drew criticism as the vehicles could be penetrated even by the anti-tank rifles issued to Finnish infantry units. The T-40A had more armour (the hull outline was altered) and the T-40S, intended to be introduced in 1942 was even better protected. However, the extra weight involved meant that the T-40S lost its amphibious capability. As a result the whole light amphibious tank concept was scrapped and production concentrated on the Light Tank T-60 (next entry). Most T-40s in service at the time of the German invasion were destroyed or captured in the vast encirclements of 1941.

Specification

Crew: 2
Combat weight: 5,900 kg (13,009 lb)
Length: 4.11 m (13.48 ft)
Width: 2.33 m (7.64 ft)
Height: 1.95 m (6.40 ft)
Engine power: 70 hp (52.22 kW)
Max speed: 44 km/h (27.32 mph)
Range: 360 km (223.56 miles)
Main gun: 12.7 mm (0.50 in) machine gun
Armour: 7 mm (0.28 in) to 14 mm (0.55 in)

T-60 Light Tank

T-60s took part in the early Russian offensives of 1942

The T-60 Light Tank was created by removing the amphibious equipment from the T-40. As it was not intended to float, the front hull and turret armour was increased to a maximum of 20 mm (0.79 in). It was armed with a 20 mm cannon with a coaxial 7.62 mm (0.30 in) machine gun. The offset turret was a new type copied from a Swedish design.

The T-60 retained the same layout as the T-40 and also used as many automobile components as possible. It was a serviceable little tank, issued to Red Army reconnaissance units and deployed on all fronts, often operating in deep snow when no other vehicles could travel. Well over 6,000 T-60s were produced, including the T-60A which had increased armour and, to speed production, solid steel wheels in place of the former spoked components.

The T-60 was still considered to be too lightly armoured and the 20 mm cannon armament inadequate. A replacement, the T-70 was introduced from 1942, but many T-60s soldiered on. They were highly reliable, even in the grim conditions prevailing on the Russian front.

Some were modified to carry Katyuscha rocket launcher rails, others became anti-tank gun tractors - the gun crews travelled on top of the hull, hanging on as well as they could. The German Army also employed captured Russian T-60s as artillery tractors, towing both field guns and anti-tank weapons. Others served as supply carriers, usually with their turrets removed.

Specification

Crew: 2
Combat weight: 5,800 kg (12,789 lb)
Length: 4.1 m (13.45 ft)
Width: 2.3 m (7.55 ft)
Height: 1.74 m (5.70 ft)
Engine power: 70 hp (52.22 kW)
Max speed: 44 km/h (27.32 mph)
Range: 450 km (279.45 miles)
Main gun: 20 mm (0.79 in)
Armour: 7 mm (0.28 in) to 20 mm (0.79 in)

T-70 Light Tank

A T-70 light tank leading a column of SU-76 assault guns

Although the T-70 Light Tank, which was first rolled out in January 1942, was based on the chassis of the earlier T-60 (previous entry) it was actually a major redesign. The single engine of the T-60 was changed to two GAZ truck engines with the transmission driving the front drive wheels. The T-60 suspension was strengthened to accommodate the extra weight of armour and the turret which now housed a 45 mm

(1.78 in) anti-tank gun. From March 1942 to October 1943 some 8,226 T-70s were manufactured and the T-70 served until the end of the war. A sound but unexceptional reconnaissance vehicle, its combat value was limited because the commander had to operate the gun as well as command the vehicle.

The T-70A had extra armour and slightly up-rated engines but these did little to alter the overall performance. After 1943 production switched from T-70 light tanks to adapting the same chassis for the SU-76 assault gun (next entry) but when the war ended many T-70s were still in service. Captured T-70s were pressed into service by the Germans and their allies. With their turrets removed they were employed as anti-tank or light field gun tractors.

Specification

Crew: 2
Combat weight: 9,200 kg (20,286 lb)
Length: 4.29 m (14.08 ft)
Width: 2.32 m (7.61 ft)
Height: 2.04 m (6.69 ft)
Engine power: 2 x 70 hp (52.22 kW)
Max speed: 45 km/h (27.95 mph)
Range: 360 km (223.56 miles)
Main gun: 45 mm (1.78 in)
Armour: 10 mm (0.39 in) to 60 mm (2.36 in)

SU-76M Light Self-propelled Gun

A typical combat shot of a SU-76M Light Self-propelled Gun

The first Soviet self-propelled gun combined a 76.2 mm (3 in) ZiS-3 field gun with a lengthened T-70 light tank chassis (previous entry). The gun was mounted in an armoured box. The chassis was fitted with an extra road wheel. Known as the SU-76M Light Mechanised Gun, it went on to be the most numerous of all World War II Soviet armoured vehicles, apart from the T-34 tank series.

Mass production began in early 1943 although few were in action before 1944. Early examples had mechanical troubles due to drive arrangements that had already proved unworkable on early T-70 light tanks.

Although intended to double as a tank destroyer, and issued with limited quantities of armour-piercing ammunition, the SU-76M primarily served as an assault gun. Its gun could fire the same ammunition as other Soviet 76.2 mm (3 in) field guns and a high-velocity solid shot was developed to penetrate enemy armour.

Despite its widespread employment the SU-76M was never a popular vehicle with its crews. The armoured superstructure was open at the top and rear so the crew were vulnerable to attack and had no protection against the elements. The SU-76M went on to a long post-war career, serving with many nations until well into the late 1980s.

Specification

Crew: 4
Combat weight: 10,200 kg (22,491 lb)
Length: 5 m (16.41 ft)
Width: 2.7 m (8.86 ft)
Height: 2.1 m (6.89 ft)
Engine power: 2 x 85 hp (63.41 kW)
Max speed: 45 km/h (27.95 mph)
Range: 320 km (198.72 miles)
Main gun: 76.2 mm (3 in)
Armour: 10 mm (0.39 in) to 35 mm (1.38 in)

T-26 Light Infantry Tank

T-26 Light Infantry Tank, the Soviet version of the Vickers 6-ton tank

The T-26 light tank was an unashamed copy of the Vickers 6-ton tank, a commercial model produced during the late 1920s and sent to the Soviet Union in 1930. The T-26A had twin 7.62 mm (0.30 in) machine gun turrets (late variants featured heavy machine guns) while the later T-26B had a single turret mounting a 37 mm (1.46 in) high-velocity gun. The final model, the T-26S Model 1937 or T-26C, had a 45 mm (1.78 in) gun, heavier armour, a rounded-off welded hull and a revised turret.

Not all T-26s were gun tanks. The OT-26 was a flame-thrower variant of the T-26A while the OT-130 was based on the T-26B; the OT-133 was the T-26S equivalent. The most extreme T-26 variant was a glider produced by strapping wings and a tail unit directly onto a T-26! Its service career, if any, is uncertain. Bulky pontoons were designed to provide limited amphibious capability.

The T-26 series provided a high proportion of Soviet tank strength in 1941 when they took part in the Soviet invasion of Iran (in co-operation with British troops). They had seen action already, in the Spanish Civil War and during the Russian attack on Finland in 1940. Most were lost in battle against the Germans in 1941-2. T-26s captured by the Germans were usually impressed as infantry support tanks while an undefined number were converted as Panzerjager carrying French 75 mm (2.96 in) anti-tank guns.

Specification - T26S

Crew: 3
Combat weight: 10,500 kg (23,152 lb)
Length: 4.88 m (16.01 ft)
Width: 3.41 m (11.19 ft)
Height: 2.41 m (7.91 ft)
Engine power: 91 hp (67.89 kW)
Max speed: 30 km/h (18.63 mph)
Range: 225 km (139.73 miles)
Main gun: 45 mm (1.78 in)
Armour: 6 mm (0.24 in) to 25 mm (0.98 in)

BT-5 Fast Tank

A BT-5 fitted with its distinctive radio aerial rigged.

The BT 'Fast Tank' series was based on the American Christie tank with its torsion bar suspension and the ability to run on tracks or large road wheels. The BT-1 was a trials vehicle only while the BT-2 was extensively tested before being released for production in early 1932. Then came the BT-3 and BT-4, neither of which was adopted.

The BT-5 appeared in 1935. It was fast, had a good power-to-weight ratio and was well-armed with a 45 mm (1.78 in) high-velocity gun and coaxial machine gun. Powered by an engine derived from an aircraft unit, the BT-5 could cross rough terrain at speed on its wide tracks, an ability often demonstrated during pre-war manoeuvres. Riveted construction was employed throughout.

A new factory at Kharkov built thousands of BT-5s and they formed a significant part of the Soviet tank arm in 1941. About 50 went to Spain to fight on the Republican side in the Civil War and the lessons learned then were incorporated into later models. There were several variants, including a command version with antennae frames around the turret. In 1941 the BT-5 proved to be under-gunned and under-armoured, but it had laid the foundations for the Soviet tanks which were to win the war.

Specification
Crew: 3
Combat weight: 11,500 kg (25,357 lb)
Length: 5.58 m (18.31 ft)
Width: 2.23 m (7.32 ft)
Height: 2.25 m (7.38 ft)
Engine power: 400 hp (298.4 kW)
Max speed: 72 km/h (44.71 mph)
Range: 200 km (124.2 miles)
Main gun: 45 mm (1.78 in)
Armour: 6 mm (0.24 in) to 13 mm (0.51 in)

BT-7 Fast Tank

BT-7 fast tanks preparing for a parade in Moscow, 1940

The BT-7 was a development of the BT-5 (previous entry), incorporating lessons learnt in the Khalin Gol battles against the Japanese. It retained the same automotive features but had an all-welded hull with sloping front plates to enhance protection, a welded turret with sloping sides, and new track. The BT-5's little-used capability to travel on the road wheels rather than the tracks was removed.

The BT-7 had several variants, including command tanks which, by 1941, often had whip aerials in place of frame antennae. The OP-7 was a flame-thrower tank while the BT-7A was an artillery support variant with a short-barrelled 76.2 mm (3 in) howitzer in an enlarged turret. The final BT-8 (or BT-7M) version had a diesel engine and a machine gun to the back of the turret. Only 706 BT-8s were produced compared to the 7,000 or so of all other BT models.

BT-7s featured in the Red Army's advances into Poland in the wake of the 1939 German invasion, and in Finland during the Winter War. But it was the German invasion of Russia in 1941 that saw the end of the BT series' combat career. Shown to be under-gunned and under-armoured, especially against the Panzer III and Panzer IV, it was phased out during 1942.

Specification

Crew: 3
Combat weight: approx 14,000 kg (30,870 lb)
Length: 5.66 m (18.57 ft)
Width: 2.29 m (7.51 ft)
Height: 2.42 m (7.94 ft)
Engine power: 500 hp (373 kW)
Max speed: 86 km/h (53.41 mph)
Range: 250 km (155.25 miles)
Main gun: 45 mm (1.78 in)
Armour: 6 mm (0.24 in) to 13 mm (0.51 in)

T-34/76 Medium Tank

The revolutionary T-34/76 tank showing its sloping armour, wide tracks and 76.2 mm gun

The T-34 gave the German army a nasty shock. Developed from the BT series, it had much heavier armour, wide tracks, a powerful engine (taken from the BT-8), and a hard-hitting 76.2 mm (3 in) gun. It was superior to any German tank in service in 1941.

The German invasion forced production to move from Kharkov and other centres to a vast factory complex east of the Ural mountains, known as Tankograd. From there T-34s streamed in thousands: from 1940 to 1945 the number of T-34s of all models was of the order of 40,000. Hurriedly built and poorly finished by western standards, the final product was nevertheless a superb fighting machine.

Early T-34s had a two-man, hexagonal-shaped, all-welded turret which was cramped and lacked vision devices for the commander. Early models had a short-barrelled 76.2 mm gun which was later replaced by a longer-barrelled model with even better anti-armour performance, resulting in a larger turret with a cupola for the commander.

Late production T-34/76 turrets were cast steel. To counter the T-34/76 the Germans developed the Panther but also employed as many captured examples as they could. Other variants of T-34/76 were few, other than flame-thrower tanks with the flame gun in the hull machine gun position.

Specification
Crew: 4
Combat weight: approx 26,000 kg (57,330 lb)
Length: 5.92 m (19.42 ft)
Width: 3 m (9.84 ft)
Height: 2.45 m (8.04 ft)
Engine power: 450 hp (335.7 kW)
Max speed: 55 km/h (34.16 mph)
Range: 300 km (186.3 miles)
Main gun: 76.2 mm
Armour: 15 mm (0.59 in) to 45 mm (1.78 in)

T-34/85 Medium Tank

A T-34/85 tank captured by the South Africans in Angola during the mid-1980s

The T-34/85 tank was first produced during the winter of 1943-4. It had a new 85 mm (3.35 in) gun, mounted in a cast steel turret originally developed for the KV-85 heavy tank. The rest of the tank was virtually unchanged from the original T-34/76. The enlarged turret provided space for three crew so the commander was at last free to fulfil his main function of commanding the tank. The wide tracks enabled the vehicle to traverse all types of ground, including soft mud and snow, allowing them to operate where

German tanks could not travel.

By 1944 the T-34/85 was employed as an infantry carrier as well as a tank. Hand rails welded around the hull enabled 'tank descent' infantry to travel into battle. Variants included flame-thrower versions with the flame gun in the bow machine gun position, SU self-propelled guns (q.v.), mine-roller tanks and bridgelayers. Battle-weary T-34s had their turrets removed and were then employed as recovery vehicles but were actually little more than tractors for disabled tanks.

The T-34/85 was regarded by the Germans as 'the best tank in the world'. Widely exported after the war, it saw front-line service in Africa as late as the 1980s and some are still in use in the former Yugoslavia and Albania.

Specification

Crew: 5
Combat weight: approx 32,000 kg (70,560 lb)
Length: 8.15 m (26.74 ft)
Width: 3 m (9.84 ft)
Height: 2.6 m (8.53 ft)
Engine power: 500 hp (373 kW)
Max speed: 55 km/h (34.16 mph)
Range: 360 km (223.56 miles)
Main gun: 85 mm (3.35 in)
Armour: 20 mm (0.79 in) to 90 mm (3.55 in)

SU-85 Assault Gun

The short-lived Su-85 assault gun

Whereas the SU-122 had been designed as a multi-purpose assault gun the SU-85 was intended to be a dedicated tank destroyer. Design work commenced in August 1943 and was considerably assisted by the experience gained during the development of the SU-122 (previous entry). The armoured superstructure of the SU-85 was virtually identical to that of the SU-122. However the SU-85's 85 mm (3.35 in) gun, based on an anti-aircraft gun, was mounted in an armoured ball mantlet, also employed by late production SU-122s.

By August 1944 the first 100 SU-85s were ready to enter service with the Red Army's newly-formed tank destroyer battalions. Like the German Panzerjager, the

Su-85 was handicapped by the limited traverse of its main armament. If a target moved outside its field of fire, the Su-85 often had to manoeuvre to engage it. Once involved in a mobile battle, a tank with a fully traversable turret had the advantage. More difficult to explain, the Su-85 lacked a machine gun, making it vulnerable to enemy infantry with anti-tank weapons, a danger countered by SU-85s standing well back from their targets and engaging them at long ranges.

The introduction of the T-34/85 medium tank (q.v.) with the same armament made the SU-85 obsolete. By early 1945 they had been withdrawn by the Red Army and replaced by the SU-100 (next entry). Surviving Su-85s were handed on to the armies of the Soviet satellite states in eastern Europe.

Specification

Crew: 4
Combat weight: 29,200 kg (64,386 lb)
Length: 8.15 m (26.74 ft)
Width: 3 m (9.84 ft)
Height: 2.45 m (8.04 ft)
Engine power: 500 hp (373 kW)
Max speed: 47 km/h (29.19 mph)
Range: 400 km (248.4 miles)
Main gun: 85 mm (3.35 in)
Armour: 20 mm (0.79 in) to 45 mm (1.78 in)

SU-100 Medium Mechanised Gun

The Su-100 served on long after World War II

Replacing the Su-85 in production from mid-1944, the SU-100 was identical except for its long 100 mm (3.94 in) gun, developed from a high-velocity naval gun with a formidable armour penetration performance. A raised cupola provided all-round vision devices for the commander.

It was a formidable tank-killer, even against the German Panthers and Tigers, and remained in service with most Warsaw Pact armies until the 1960s and with other Soviet-influenced armed forces as late as the 1980s. Israeli tank crew encountered them in the campaigns of both 1956 and 1967.

Although intended primarily for destroying tanks, the 100 mm gun could also fire a useful high explosive shell. As with the SU-85 there was no secondary machine gun armament, so close combat had to be avoided. One practice which was adopted as standard was to bolt spare track links over the front hull and superstructure sides to increase armour protection.

A tactical limitation for the SU-100 was that the internal ammunition stowage was limited by the length of the one-piece rounds involved. The SU-100 could accommodate only 34 100 mm rounds on racks around the fighting compartment walls, compared to 48 85 mm (3.35 in) rounds on the SU-85. This was often alleviated by carrying ammunition in cases stowed on the rear hull but this was not a practice to be encouraged during combat. It was bad enough that the SU-100s carried external fuel tanks.

Specification

Crew: 4
Combat weight: 31,600 kg (69,678 lb)
Length: 9.45 m (31.00 ft)
Width: 3 m (9.84 ft)
Height: 2.25 m (7.38 ft)
Engine power: 500 hp (373 kW)
Max speed: 48 km/h (29.8 mph)
Range: 320 km (198.72 miles)
Main gun: 100 mm (3.94 in)
Armour: 20 mm (0.79 in) to 45 mm (1.78 in)

T-35 Heavy Tank

The gigantic T-35s made an impressive show on parade

The Red Army was always a proponent of the heavy tank, and by 1930 even had a project for a 100 ton tank. That came to naught, but a less ambitious design appeared in the form of the T-32. Influenced by the British Vickers Independent (a commercial design, the details of which the Soviets took considerable pains to acquire), the T-32 was only an interim model for the full production version, the T-35 heavy tank.

Besides the main turret which housed a short 76.2 mm howitzer, there were two 37 mm (1.46 in) gun turrets and two machine gun turrets and there was another machine gun position in the front hull. The crew was no less than 11 men!

The first T-35 was ready in July 1932 and was first displayed at the annual May Day parade in 1933. Its massive dimensions and impressive-looking armament caused an international stir. What was not so apparent was that the great length of the hull made it very difficult to steer, and its armour was thin. The guns could only be fired accurately if the tank was stationary. To add to this the T-35 was a lumbering beast across country and had only a limited operational range.

Production was very slow. Only about 60 T-35s were ever made. These were stationed near Moscow and continued to feature in military parades under the baleful eye of Stalin. A handful took part in the campaign against Finland and the rest were lost in 1941, several captured intact, but out of fuel.

Specification

Crew: 11
Combat weight: approx 45,000 kg (99,225 lb)
Length: 9.72 m (31.89 ft)
Width: 3.2 m (10.50 ft)
Height: 3.43 m (11.25 ft)
Engine power: 500 hp (373 kW)
Max speed: 30 km/h (18.63 mph)
Range: 150 km (93.15 miles)
Main gun: 1 x 76.2 mm howitzer; 2 x 37 mm (1.46 in)
Armour: 11 mm (0.43 in) to 30 mm (1.18 in)

KV-1 Heavy Tank

The impressive bulk of the KV-1 heavy tank armed with a 76.2 mm gun

The T-100 heavy tank was supposed to be the successor to the T-35 (previous entry). It had two turrets, one with a 76.2 mm gun, the other with a 45 mm (1.78 in) gun, but it was an unwieldy beast and was passed over in favour of a simplified heavy tank which became the KV-1.

The first KV-1s were prone to clutch and transmission faults which took time to eliminate. The main armament was a 76.2 mm gun but, although the armour used was the thickest that could be produced at that time, it was still considered insufficient, a drawback which was overcome by adding sheets of applique armour to the front hull and turret. Later improvements included a longer-barrelled 76.2 mm gun, and the welded turret was provided with a cast steel component. KV-1s were continually up-armoured to make them virtually invulnerable to German anti-tank guns. Yet the weight of all this extra armour was not compensated for by any added power from the engine which remained unchanged.

The KV-1 remained in production until 1943. By then production was switching to the KV-85 which had an 85 mm (3.35 in) gun in a new turret, the same combination being adopted for the T-34/85 (q.v.). Only 130 KV-85s were produced.

Specification

Crew: 5
Combat weight: approx 43,000 kg (94,815 lb) (later 52,000 kg (114,660 lb))
Length: 6.68 m (21.92 ft)
Width: 3.32 m (10.89 ft)
Height: 2.71 m (8.89 ft)
Engine power: 600 hp (447.6 kW)
Max speed: 35 km/h (21.74 mph)
Range: 335 km (208.03 miles)
Main gun: 76.2 mm
Armour: 25 mm (0.98 in) to 75 mm (2.96 in) (later 130 mm (5.12 in))

KV-2 Heavy Artillery Tank

The KV-2's huge turret made it an obvious target

Almost as soon as the KV-1 heavy tank (previous entry) was conceived it was decided to develop a close support version with a heavy gun. The result was the KV-2 heavy artillery tank which initially had an enlarged turret mounting a 122 mm (4.81 in) howitzer, soon changed to a taller turret with a 152 mm (6 in) howitzer. The hull and chassis of the KV-1 was used virtually unchanged.

The KV-2 was produced in much smaller numbers than the KV-1 which was just as well. The KV-2 performed well enough in set-piece assaults on Finnish positions in 1940, but against the Germans in 1941 it was a very different story. The huge slab-sided turret with its four-man crew was so bulky that it was difficult to traverse quickly, especially on an incline. Its great weight made the KV-2 painfully slow and unstable. Its high profile drew enemy fire. Most KV-2s were lost during the first months of the German invasion in 1941 and production ceased during that year as the factories were overrun.

There was only one variant, the KV-2B. Based on the KV-1B chassis, it had wider tracks for better mobility over soft ground or snow plus a hull-mounted machine gun. A 1943 plan involved fitting an 85 mm (3.35 in) gun instead, and at least one KV-2 was used for trials with a modified 107 mm (4.22 in) naval gun which would have formed the main armament of the proposed KV-3. Both schemes were abandoned.

Specification

Crew: 6
Combat weight: approx 52,000 kg (114,660 lb)
Length: 6.8 m (22.31 ft)
Width: 3.32 m (10.89 ft)
Height: 3.28 m (10.76 ft)
Engine power: 600 hp (447.6 kW)
Max speed: 26 km/h (16.15 mph)
Range: 225 km (139.73 miles)
Main gun: 152 mm (6 in) howitzer
Armour: 30 mm (1.18 in) to 110 mm (4.33 in)

IS-1 Heavy Tank

The IS-1 was named after the Soviet leader, Iosef Stalin

The IS-1 (IS - Iosef Stalin) was never to be produced in quantity but it marked an important step in Soviet tank development, leading to what was to become the most powerful tank of the immediate post-war years. It was a new design which owed much to the previous KV series (q.v.) but with a modified hull and improvements to the suspension, transmission and power train. It was heavily armoured and had a new turret, mounting the same 85 mm (3.35 in) gun as the T-34/85 which by then (late 1943) was already on the production lines.

The need for heavier guns led to trials with 100 mm

(3.94 in) and 122 mm (4.81 in) guns, the 100 mm gun proving superior. However, the performance of the 122 mm gun was only marginally less effective and there was spare capacity to produce the 122 mm gun in quantity, as well as an established ammunition production base (which was limited for the 100 mm gun). The 122 mm weapon was selected for the new main armament, but it was never actually fitted to the IS-1. Instead all development was switched to the even more promising IS-2 (next entry).

A few IS-1s saw action in the Ukraine in early 1944, operating in the heavy breakthrough role for which they were ideally suited. The 100 or so IS-1s which were produced were later converted to IS-2 standards.

Specification

Crew: 4
Combat weight: approx 46,000 kg (101,430 lb)
Length: 8.32 m (27.30 ft)
Width: 3.25 m (10.66 ft)
Height: 2.9 m (9.51 ft)
Engine power: 600 hp (447.6 kW)
Max speed: 40 km/h (24.84 mph)
Range: 250 km (155.25 miles)
Main gun: 85 mm (3.35 in)
Armour: 30 mm (1.18 in) to 160 mm (6.30 in)

IS-2 Heavy Tank

An IS-2 moving past resting Red Army soldiers

The IS-2 heavy tank was essentially the IS-1 with the turret modified to accept the long-barrelled high-performance 122 mm (4.81 in) gun, by far the most powerful weapon to be mounted in a combat tank turret during World War II. Its only drawback was that the ammunition was of the two-part type which slowed down loading and it was so bulky that only 28 rounds could be carried.

By the time production ended about 2,250 IS-2s

had been built. The IS-2m had a better fire control system, extending the effective range of its main armament. In battle the IS-2 was able to tackle German Panthers and Tigers on equal terms. IS-2s were in action throughout the latter stages of the war, culminating in the final assault on Berlin, although by then they had been joined by small numbers of a new IS model.

This was the IS-3, essentially the IS-2 redesigned to lower the silhouette and create shot deflection surfaces around the 'frying pan' outline turret and the pointed glacis plate. The IS-3 went on to be the bugbear of Western tank designers for years after the war and can lay claim to being the most powerful of all the World War II tanks, even if its main service life was to be post-war.

Specification - IS-2

Crew: 4
Combat weight: approx 46,000 kg (101,430 lb)
Length: 9.9 m (32.48 ft)
Width: 3.09 m (10.14 ft)
Height: 2.73 m (8.96 ft)
Engine power: 600 hp (447.6 kW)
Max speed: 37 km/h (22.98 mph)
Range: 240 km (149.04 miles)
Main gun: 122 mm (4.81 in)
Armour: 30 mm (1.18 in) to 160 mm (6.30 in)

SU-122 Medium Mechanised Gun

Side view of a SU-122 assault gun showing the heavy mantlet for the 122 mm howitzer

The Su-122 was an assault gun based on the T-34 medium tank (q.v.). The turret was replaced by an armoured superstructure housing a modified 122 mm (4.81 in) field howitzer. The first production examples, designated the SU-122 Medium Mechanised Gun, were issued to the Red Army in January 1943.

The SU-122 proved to be effective as an assault gun, delivering direct fire during attacks on strongly defended positions. Unfortunately the anti-armour performance of the 122 mm (4.81 in) howitzer was less than expected. A shaped charge HEAT (high explosive anti-tank) projectile was developed but it was only accurate at short range and armour penetration was still disapointing. Replacing the howitzer with a 122 mm field gun overloaded the chassis, so the SU-85 (next entry) was produced instead. The SU-122 soldiered on, sometimes with the armed forces of Soviet allies, until the war ended.

In place of the usual awkward-looking armoured mantlet used on most SU-122s, late production models had a ball mantlet similar to that adopted on the SU-85.

The Germans were happy to employ as many SU-122s as they could capture, usually banded into special units and liberally decorated with German markings. Since it was based on the T-34, which had also been captured in large numbers, there were usually sufficient spare parts to keep the Su-122s operational for a reasonable period. Ammunition presented little problem either, as the 122 mm field gun was already used by the Germans as well.

Specification

Crew: 5
Combat weight: 30,900 kg (68,134 lb)
Length: 6.95 m (22.80 ft)
Width: 3 m (9.84 ft)
Height: 2.32 m (7.61 ft)
Engine power: 500 hp (373 kW)
Max speed: 55 km/h (34.16 mph)
Range: 300 km (186.3 miles)
Main gun: 122 mm (4.81 in) howitzer
Armour: 20 mm (0.79 in) to 45 mm (1.78 in)

ISU-152 Heavy Mechanised Gun

A highly impressive monster, the ISU-152 Heavy Mechanised Gun with its 152 mm howitzer

The first Soviet tracked carriage to mount a 152 mm (6 in) gun-howitzer was the SU-152 dating from mid-1943. Based on the KV-1 chassis, it arrived just in time to take part in the battle of Kursk and proved itself able to engage all German armour. Its massive 43.5 kg (95.92 lb) projectile could destroy the heaviest German tanks.

When the KV-1 production line closed, it was decided to fit a similar weapon on the IS series chassis (q.v.). This became the ISU-152 and was every bit as successful as the earlier model. Two types of gun were employed: the 152 mm (6 in) gun-howitzer and the

122 mm (4.81 in) gun as used on the IS-2 tank - this became the ISU-122. Since the IS-2 tank carried the same gun, only a few ISU-122s were built, but they served until the end of the war.

ISU-152s took part in the Battle for Berlin. With no secondary armament other than a 12.7 mm (0.50 in) machine gun over a roof hatch, they had to rely on infantry support. Their powerful armament enabled the ISU-152s to destroy German strongpoints holding up the Russian advance.

Only 20 rounds could be carried inside, so more were often stowed on the rear decking, as were extra fuel tanks to increase the operational range. The ISU-152 remained in production after the war ended and was developed (post war) into the ISU-130 with a 130 mm (5.12 in) gun.

Specification

Crew: 5
Combat weight: approx 46,000 kg (101,430 lb)
Length: 9.18 m (30.12 ft)
Width: 3.07 m (10.07 ft)
Height: 2.48 m (8.14 ft)
Engine power: 600 hp (447.6 kW)
Max speed: 37 km/h (22.98 mph)
Range: 220 km (136.62 miles)
Main gun: 152 mm (6 in) gun-howitzer
Armour: 30 mm (1.18 in) to 90 mm (3.55 in)

Light Tank Mark VIB

A Vickers Light Tank Mark IVA armed with 2 machine guns

The Light Tank Mark VIB was the most numerous of a series of British light tanks which began in 1931. The Light Tank Mark 1 was a development of earlier experimental designs which could be traced back to the series of Carden-Loyd tankettes. Early Light Tank marks had two-man crews, increased to three with the Mark V. Armament was limited to one 0.50-inch (12.7 mm) and one 0.303-inch (7.7 mm) machine gun on the Mark VIB - a Mark VIC had one 15 mm (0.59 in) and one 7.92 mm (0.31 in) machine gun but was produced in limited numbers only. Armoured protection was very limited but the Light Tanks were agile and fairly reliable.

The Light Tanks were sent to France in 1940. When the Germans attacked in May they proved of limited combat value, even in the reconnaissance role. Their standard armament was no match for the German tanks. Most in France were either destroyed in action or simply abandoned. The same harsh lessons were re-learned in North Africa and during operations in the Lebanon in 1941 although against the Italians in 1940, the speed of the Light Tanks did prove useful. After 1942 the Light Tanks were withdrawn as soon as possible and replaced by Stuarts (q.v.). A few Light Tanks of various marks were still being used for training as late as 1943.

Specification - Mark IVB

Crew: 3
Engine power: 88 hp (65.65 kW)
Combat weight: 4,900 kg (10,804 lb)
Max speed: 56 km/h (34.78 mph)
Length: 3.95 m (12.96 ft)
Range: 200 km (124.2 miles)
Width: 2.057 m (6.75 ft)
Main gun: 0.50-inch (12.7 mm)
Height: 2.22 m (7.28 ft)
Armour: 4 mm (0.16 in) to 14 mm (0.55 in)

Tetrarch

The Tetrarch was flown into battle with British airborne forces during the crossing of the Rhine in 1945

The Tetrarch was the United Kingdom's first, and only, airborne tank. Originally known as the Light Tank Mark VII, it was to have been the follow-on to the Mark VI (previous entry). By the time the weaknesses of British light tanks had been exposed, the Mark VII was already well advanced. Started as a private venture by Vickers in 1938, it was taken over by the War Office and limited production began in July 1940. At first there seemed to be little prospect of a service career even if it did have a better all-round layout than the earlier Light Tanks, as well as a 2-pounder (40 mm) main gun. A few saw service during the take-over of Madagascar in May 1942 and during that same year a batch was sent to the Soviet Union.

That would have been the end of the Mark VII but a new role then beckoned. The British Army was forming airborne forces and a light tank able to support the paratroops and glider-borne soldiers seemed to be a good idea. Renamed Tetrarch, the design needed few modifications, although on some vehicles the 2-pounder (40 mm) gun was replaced by a close support 3-inch (76.2 mm) howitzer. The Hamilcar glider, the intended airborne transporter, was virtually designed around the dimensions of the Tetrarch. The Tetrarch was borne into action on 6 June 1944 and again during the Rhine crossings in March 1945.

Specification

Crew: 3
Engine power: 165 bhp
Combat weight: 7,620 kg (16,802 lb)
Max speed: 64 km/h (39.74 mph)
Length: 4.305 m (14.12 ft)
Range: 225 km (139.73 miles)
Width: 2.31 m (7.58 ft)
Main gun: 2-pounder (40 mm)
Height: 2.12 m (6.96 ft)
Armour: 4 mm (0.16 in) to 14 mm (0.55 in)

Cruiser Mark 1

Cruiser Mark 1 tanks leading a procession of early Cruiser tanks on exercise in the UK

The first of the British Cruiser tanks, the Mark 1 was built by Vickers Armstrong as the A9E1. Designed in 1936, it was a considerable advance over other British tanks of the era such as the Vickers Medium series. It even had a hydraulic powered traversing mechanism for the main turret. The three-wheel bogie suspension was so successful it was adopted for the Valentine infantry tank (q.v.). However, there were carry-overs from earlier Vickers models such as the two 0.303-inch (7.7 mm) machine gun turrets on the front hull; when these were manned the Cruiser Mark 1 required a crew of six. The main armament was a 2-pounder (40 mm) gun in the turret which was a hard-hitting weapon by 1930s standards.

By 1937 the Mark 1 was in limited production and a batch of 125 was ordered. A few were fitted with a low-velocity 3.7-inch (94 mm) close support howitzer in place of the 2-pounder (40 mm) and were renamed Cruiser Mark 1CS (Close Support).

In the 1940 campaign the Mark 1 was found wanting in protection, firepower and speed. Its off-road performance was poor, as was its mechanical reliability. All those sent to France were lost. Another batch was lost in Greece the following year. In North Africa Cruiser Mark 1s had to soldier on until well into 1942. A few captured examples were briefly used by the Germans.

Specification

Crew: 6
Engine power: 150 hp (111.9 kW)
Combat weight: 13,040 kg (28,753 lb)
Max speed: 40 km/h (24.84 mph)
Length: 5.79 m (19.00 ft)
Range: 240 km (149.04 miles)
Width: 2.5 m (8.20 ft)
Main gun: 2-pounder (40 mm)
Height: 2.654 m (8.70 ft)
Armour: 6 mm (0.24 in) to 14 mm (0.55 in)

Cruiser Mark IIA

Loading a Cruiser Mark II onto a railway flatcar

Designed in 1936, the Vickers Armstrong Cruiser Mark II followed on from the Mark 1. Intended to operate in an infantry support tank, it had heavier armour. It was not actually used in this role, but the Mark II was often referred to at the time as a 'heavy cruiser'; the design designation was A10.

The most noticeable change from the Mark 1 was the removal of the twin auxiliary machine gun turrets. After some debate, it was finally accepted that these were of limited combat value and wasted valuable manpower. The single coaxial machine gun retained was a 7.92 mm (0.31 in) BESA machine gun, a calibre and type adopted only by the Royal Armoured

Corps. As with the Mark 1 series there was a Cruiser Mark IICS armed with a 3.7-inch (94 mm) close support howitzer. There was little difference between the two marks apart from armament; they even shared the same AEC engine. 170 of the Cruiser Mark IIA main production model were ordered in 1938.

In the event the Cruiser Mark IIA fell between two stools. It lacked sufficient armour for the infantry support role, yet it was too slow to be a true Cruiser tank. The first examples were delivered during December 1939 and a batch was sent to France, only to vanish in May 1940. More were sent to North Africa where they proved to be mechanically more reliable than the Mark 1s; they served there until 1942 when sufficient Crusaders and Lee/Grants had been sent to the Eighth Army.

Specification

Crew: 5
Engine power: 150 hp (111.9 kW)
Combat weight: 14,390 kg (31,729 lb)
Max speed: 26 km/h (16.15 mph)
Length: 5.59 m (18.34 ft)
Range: 160 km (99.36 miles)
Width: 2.527 m (8.29 ft)
Main gun: 2-pounder (40 mm)
Height: 2.654 m (8.70 ft)
Armour: 4 mm (0.16 in) to 30 mm (1.18 in)

Cruiser Marks III and IV

A wrecked Cruiser Mark IV in the aftermath of the Battle of France, 1940

A radical change was introduced into British tank design with the Cruiser Mark III. It was the first British tank to adopt the Christie suspension which was not only able to absorb considerable punishment but enabled the tank to travel at much higher speeds. This change followed a demonstration of Christie-equipped Soviet BT-5 tanks (q.v.) during a 1936 visit by British observers.

The Nuffield Organisation were requested to produce prototypes, designated the A13, which were ready in 1938. By 1940 Cruiser Mark IIIs were in service in France, but only 65 Mark IIIs were built as its shortcomings were soon apparent. They were joined in France by the Cruiser Mark IV, which had extra armour and a larger turret. This became the most important British tank (numerically) of the early war years, bearing the brunt of the fighting in France and the first desert campaigns. Production ceased in 1941 after 655 had been delivered

The Cruiser Marks III and IV were not a success. Their armour was too light, even against German tank guns. Cruiser tanks struck by the powerful anti-tank guns used by the Germans in North Africa seldom survived. The Liberty engine and its transmission were unreliable, and the 2-pounder (40 mm) main armament was inadequate by 1941.

Specification - Mark IV

Crew: 4
Engine power: 340 hp (253.64 kW)
Combat weight: 15,000 kg (33,075 lb)
Max speed: 48 km/h (29.8 mph)
Length: 6.02 m (19.75 ft)
Range: 150 km (93.15 miles)
Width: 2.54 m (8.33 ft)
Main gun: 2-pounder (40 mm)
Height: 2.59 m (8.48 ft)
Armour: 6 mm (0.24 in) to 30 mm (1.18 in)

Covenanter

A row of Covenanter tanks lined up for inspection

Despite the disappointing performance of the Cruiser Marks III and IV (previous entry) the Christie suspension was deemed a success and was adopted for the rest of the Cruiser tank series. The Cruiser Mark V, or Covenanter, was basically an A13 reworked in an attempt to overcome many of the mechanical difficulties to which that design was prone. The work was carried out by the London Midland and Scottish Railway Company.

The design featured a new and highly successful

steering system. To achieve a lower silhouette it was
fitted with a Meadows-Fiat 'flat-12' engine in the rear
with the radiator louvres on the front of the hull. This
produced the required height reduction but created a
new problem: the engine cooling system did not
work. Four different marks were built in an attempt
to solve the cooling problems but they were never
overcome.

In the rush to production there had been no
thorough trials and the cooling problems were not
revealed until the Covenanter was coming off the
lines in quantity. A total of 1,771 were built but they
were never committed to battle. Covenanters served
as training vehicles, the last of them being withdrawn
during 1943.

Specification

Crew: 4
Engine power: 300 hp (223.8 kW)
Combat weight: 18,300 kg (40,351 lb)
Max speed: 50 km/h (31.05 mph)
Length: 5.8 m (19.03 ft)
Range: 160 km (99.36 miles)
Width: 2.61 m (8.56 ft)
Main gun: 2-pounder (40 mm)
Height: 2.23 m (7.31 ft)
Armour: 7 mm (0.28 in) to 40 mm (1.58 in)

Crusader

A Crusader knocked out in the North African desert

While the sorry saga of the Covenanter unfolded, Nuffields were working on their own A13 improvement programme. They retained the Liberty engine and lengthened the Christie suspension by an extra road wheel to create the A15, Cruiser Mark VI, or Crusader. The first of them entered service in 1941 and were shipped off to North Africa.

Another British tank rushed into production without thorough testing, the Crusader Mark I was mechanically unreliable. Its 2-pounder (40 mm) was well outclassed by 1941 and for some reason it had a machine gun in a small turret on the front hull. The latter was removed on the Crusader Mark II but only on the Mark III was the main armament at last up-gunned to a 57 mm 6-pounder which allowed it to fight German armour on fairly equal terms. There was also a Crusader IICS armed with a 3-inch (76.2 mm) close support howitzer.

The Crusader formed the main equipment of the armoured divisions fighting the Desert Campaigns; over 5,300 were produced. Once withdrawn, Crusaders were used for a variety of special purposes including observation post and command tank, ARVs, and mine-clearing. Turretless Crusaders were used as 17-pounder (76.2 mm) anti-tank gun tractors and air defence models with either a single 40 mm (1.58 in) Bofors Gun or twin or triple 20 mm (0.79 in) cannon.

Specification - Mark III

Crew: 5
Engine power: 340 bhp
Combat weight: 20,085 kg (44,287 lb)
Max speed: 43 km/h (26.70 mph)
Length: 5.994 m (19.67 ft)
Range: 200 km (124.2 miles)
Width: 2.642 m (8.67 ft)
Main gun: 6-pounder (57 mm)
Height: 2.235 m (7.33 ft)
Armour: 7 mm (0.28 in) to 49 mm (1.93 in)

Cavalier

The Cavalier was rushed into production without testing

The service career of the Crusader series (previous entry) was closely monitored by Nuffields. By 1941 they acknowledged that a completely new model was required: This was to be the Cromwell (q.v.). But it was also realised that the engines for such a model would not be available in quantity for some time. In the meantime the British Army was desperately short of tanks. Nuffield's interim response was to update the Crusader.

The result was known initially as the A24. Yet again the demand was such that 500 were ordered 'off the drawing board' and rushed into production as the Cruiser Mark VII or Cavalier. Although up-gunned and up-armoured, the old Liberty engine was retained. By 1942 it was obvious that this engine and general drive train were prone to trouble. As a result the Cavalier was never sent into action and was used as yet another training tank. Some saw action as artillery observation tanks while others became ARVs (Armoured Recovery Vehicles).

The Cavalier marked the end of the 'rush to production' syndrome of the early war years. By 1942 American tanks were available in sufficient numbers to meet the United Kingdom's immediate needs. With more time for testing before new tanks went into production, the general reliability level of British tanks rose accordingly. By the end of World War II, British tank designs were again of high quality.

Specification

Crew: 5
Engine power: 410 hp (305.86 kW)
Combat weight: 26,950 kg (59,424 lb)
Max speed: 39 km/h (24.22 mph)
Length: 6.35 m (20.83 ft)
Range: 265 km (164.57 miles)
Width: 2.88 m (9.45 ft)
Main gun: 6-pounder (57 mm)
Height: 2.428 m (7.97 ft)
Armour: 20 mm (0.79 in) to 76 mm (2.99 in)

Centaur

A Centaur heavy cruiser tank armed with a 75 mm gun

The Centaur was a hybrid, being a Cromwell (next entry) with a Liberty engine. It was produced when there was no prospect of the Cromwell being placed in production due to a lack of the required Meteor engines. The interim Cruiser Mark VIII, Centaur, was fitted with the Liberty engine by Morris Engines, under the design designation A27L. Many Centaurs were later retrofitted with Meteor engines, and designated the Cromwell X.

The first Centaurs appeared during early 1942 and were produced by Leyland Motors. For an interim vehicle, the Centaur was to have a wide ranging career. Although most were used for training the Centaur Mark IV was armed with a 95 mm (3.75 in) howitzer and issued to the Royal Marines to provide close fire support during the Normandy landings of June 1944. They were supposed be used only in the initial stages of the landings, but they were pressed into service for weeks afterwards. The Centaur remains one of the few tanks ever used by the Royal Marines.

The original Centaur Mark 1 was armed with a 6-pounder (57 mm) gun. The Mark III had a 75 mm (2.96 in) gun. The Centaur Kangaroo was a turretless armoured personnel carrier while the Centaur OP retained its turret but was fitted with a dummy gun to disguise its artillery observation role. There were two marks of Centaur AA tanks armed with 20 mm (0.79 in) cannon and some Centaurs were converted for the armoured recovery role.

Specification - Mark IV

Crew: 5
Engine power: 395 hp (294.67 kW)
Combat weight: 28,875 kg (63,669 lb)
Max speed: 43 km/h (26.70 mph)
Length: 6.35 m (20.83 ft)
Range: 265 km (164.57 miles)
Width: 2.896 m (9.50 ft)
Main gun: 95 mm (3.75 in) howitzer
Height: 2.49 m (8.17 ft)
Armour: 20 mm (0.79 in) to 76 mm (2.99 in)

Cromwell

Cromwell tanks advancing through Holland, early 1945

The Cromwell was built on the lessons learned with previous British tank designs. It was the first wartime British tank to have a reliable engine, the Rolls Royce Meteor, a derivation of the Merlin aircraft engine. Demand for the latter delayed its use in tanks, leading to the Centaur (previous entry). The Meteor-powered tank was the Cruiser Mark VIII Cromwell, the A27M. Production was carried out by Leyland

Motors from mid-1943 onwards.

The Cromwell had a well-tried suspension, a powerful engine with capacity for development, and good protection. The first models were armed with a 6-pounder (57 mm) gun although this was uprated to a 75 mm (2.96 in) gun as soon as they became available. The Cromwell Mark VI, was armed with a 95 mm (3.75 in) close support howitzer. Some Centaurs were re-engined to become Cromwell Xs. Some Cromwells were converted to serve as ARVs.

At last British tank crews were on a par with their German opponents. Cromwells were not ready for action until May 1944 after which they became the most important, in both numerical and quality terms, of the British tanks and remained in service until years after the war ended.

Specification - Mark IV

Crew: 5
Engine power: 600 hp (447.6 kW)
Combat weight: 27,970 kg (61,673 lb)
Max speed: 51 km/h (31.67 mph)
Length: 6.35 m (20.83 ft)
Range: 278 km (172.64 miles)
Width: 2.91 m (9.55 ft)
Main gun: 75 mm (2.96 in)
Height: 2.49 m (8.17 ft)
Armour: 8 mm (0.32 in) to 76 mm (2.99 in)

Challenger

The Challenger's huge turret mounted a 17-pounder anti-tank gun

It was soon clear that a version of the Centaur/Cromwell able to carry the 17-pounder (76.2 mm) anti-tank gun would be necessary to counter the new German heavy tanks. The result was the A30 or Challenger. Design work began in May 1942 but the 200 ordered were not ready until mid 1944. Even then they could not take part in the Normandy landings as no provision had been made for wading equipment.

The Challenger had a lengthened hull with an extra road wheel and a widened centre section for the bigger turret ring needed for the gun. The long 17-pounder gun demanded a taller turret as it had to accommodate an extra loader to handle the large rounds involved. To make room for more ammunition, the hull machine gun was deleted; even then, it only carried 42 rounds. The design emerged overweight so some armour had to be removed, rendering the Challenger unable to operate in close association with Cromwells, as had originally been intended.

By the time it was ready, the Challenger was rather redundant as the Sherman Firefly (q.v.) was already in widespread use in the role for which it was intended.

Specification

Crew: 5
Engine power: 600 hp (447.6 kW)
Combat weight: 33,050 kg (72,875 lb)
Max speed: 24 km/h (14.90 mph)
Length: 8.147 m (26.73 ft)
Range: 193 km (119.85 miles)
Width: 2.9 m (9.51 ft)
Main gun: 17-pounder (76.2 mm)
Height: 2.775 m (9.10 ft)
Armour: 20 mm (0.79 in) to 102 mm (4.02 in)

Comet

A Comet tank preserved in a South African museum

Almost as soon as the first of the Centaur/Cromwell tanks were completed, thoughts were given to a heavier armament. Since the prime candidate, the 17-pounder (76.2 mm) anti-tank gun required a wider turret ring, Vickers-Armstrong developed a version of the gun firing a reduced propellant charge. This gun fired the same projectiles and its performance was only slightly reduced. It was known as the 77 mm (3.03 in) for logistic differentiation.

The tank for which the 77 mm gun was intended was supposed to be an enhanced Cromwell but by the time all the design changes had been embodied there were over 60 per cent new components, resulting in an A34 designation and a new name of Comet.

The first were delivered in December 1944 and took some part in the latter stages of the war, immediately proving themselves both battle-worthy and reliable. After 1945 Comets remained a feature of British Army tank parks until 1960 and served with many other armies well after then. The Comet was the best all-round British tank of the war years.

The Comet had a slightly wider all-welded hull than the Cromwell due to the turret ring diameter which could not be reduced beyond a certain limit. There were also changes to the Christie suspension system which had remained basically the same since the original Cruiser tank Mark III.

Specification
Crew: 5
Engine power: 600 hp (447.6 kW)
Combat weight: 35,775 kg (78,883 lb)
Max speed: 47 km/h (29.19 mph)
Length: 7.65 m (25.10 ft)
Range: 198 km (122.96 miles)
Width: 3.05 m (10.00 ft)
Main gun: 77 mm (3.03 in) (actual calibre 76.2 mm (3 in))
Height: 2.68 m (8.79 ft)
Armour: 25 mm (0.98 in) to 102 mm (4.02 in)

Matilda I

The little Matilda I, first of the British Army's Infantry tanks

The Matilda I was the first of the so-called Infantry tanks, intended to provide close support for foot soldiers. Since they were supposed to operate at the same pace as a walking man they had no need for speed and the emphasis was placed on crew protection.

The initial requirement for two types of infantry tank, one small and one heavy, was made in 1934 and the first prototype of the smaller of the two was ready in 1936. The project had been given the code name Matilda (after a cartoon duck of the period) and the name stuck, but officially the design was known as the A11, or Infantry Tank Mark I.

The small Matilda was designed with two things in mind, low cost and quick production. It had only a two-man crew, its Ford V-8 engine gave a top speed of only 12.8 km/h (7.95 mph), and it was only armed with a 0.303-inch (7.7 mm) machine gun (a few had a 0.50-inch (12.7 mm) weapon).

Armour protection was excellent and overall reliability was good but the design was obsolete by 1939. Production of the Matilda I ceased after 139 had been built. Two battalions of Matilda Is were sent to France and all were lost in 1940. It might have been difficult to knock out, but its machine gun armament was useless against German armour. For once the Germans made no use of captured examples.

Specification

Crew: 2
Engine power: 70 hp (52.22 kW)
Combat weight: 11,185 kg (24,662 lb)
Max speed: 12.8 km/h (7.95 mph)
Length: 4.85 m (15.91 ft)
Range: 129 km (80.11 miles)
Width: 2.286 m (7.50 ft)
Main gun: 0.303-inch (7.7 mm) machine gun
Height: 1.867 m (6.13 ft)
Armour: 10 mm (0.39 in) to 60 mm (2.36 in)

Matilda

A Matilda infantry tank during training in Southern England

The A12, or Infantry Tank Mark II followed the Matilda I (previous entry) The first was delivered in 1938 but old-fashioned manufacturing techniques meant that each tank had to be built by skilled craftsmen. Only a few were in service by 1940. Soon known simply as Matilda, the new infantry tank was a great success. Unlike other British tanks, it was well-armoured and reliable, mainly due to the use of commercial components (such as two AEC bus engines) wherever possible. Its armour was proof against most German anti-tank weapons of the early war years, but the Matilda was very slow (only 24 km/h (14.9 mph) on roads) even after the original AEC engines had been replaced by Leyland units on the Mark IIA*.

The small turret ring diameter prevented it carrying anything more powerful than the 2-pounder (40 mm) gun. By the time production ceased in 1943, 2,987 had been built. After sterling service in North Africa most had already been withdrawn and diverted to special purpose roles including mine flail tanks, Canal Defence Lights to produce 'artificial moonlight' and assault bridges. The Australians produced a flame-thrower version, called the 'Frog'. Many Matildas were also donated to the Soviet Union.

Specification - Mark III

Crew: 4
Engine power: 2 x 190 hp (141.74 kW)
Combat weight: 26,950 kg (59,424 lb)
Max speed: 24 km/h (14.90 mph)
Length: 5.613 m (18.42 ft)
Range: 255 km (158.36 miles)
Width: 2.59 m (8.50 ft)
Main gun: 2-pounder (40 mm)
Height: 2.515 m (8.25 ft)
Armour: 20 mm (0.79 in) to 78 mm (3.07 in)

Valentine

A Valentine infantry tank on exercise with Polish Army troops in England

The Valentine, or Infantry Tank Mark III, was a Vickers development based on the Cruiser Tank Marks I and II but with heavier armour. It was ordered 'off the drawing board' and the prototype was ready on 14 February 1940, hence the name Valentine. By the time production ceased in 1944, 8,275 had been manufactured in the UK and Canada, a higher total than for any other British tank. Most Canadian examples were shipped direct to the

Soviet Union.

The initial armament of a single turret-mounted 2-pounder (40 mm) gun was replaced by a 6-pounder (57 mm) and finally a 75 mm (2.96 in) gun - there were 11 marks of Valentine tank. At first, AEC petrol engines were used but later models had diesels. Following the usual spate of 'into-service' troubles Valentines proved to be popular and reliable vehicles, even if they were difficult to drive.

Obsolete in Europe by 1943 (many soldiered on in the Far East) the Valentine was diverted to special roles. Important variants included the Bishop (next entry) and the Archer (q.v.). Valentines were used for trials involving assault bridges (some were used in Burma), mine rollers, flame-throwers, Canal Defence Lights and the Duplex Drive swimming tank systems.

Specification - Mark II

Crew: 3
Engine power: 131 hp (97.73 kW)
Combat weight: 17,700 kg (39,028 lb)
Max speed: 24 km/h (14.90 mph)
Length: 5.41 m (17.75 ft)
Range: 145 km (90.05 miles)
Width: 2.63 m (8.63 ft)
Main gun: 2-pounder (40 mm)
Height: 2.273 m (7.46 ft)
Armour: 8 mm (0.32 in) to 65 mm (2.56 in)

Bishop Self-propelled Gun

The Bishop was a hasty conversion of a Valentine tank

The British Army carried out trials with self-propelled guns between the wars but it was only when the Afrika Korps started to use them against the British in North Africa that one was actually put into production. The Bishop was a hasty conversion of a Valentine infantry tank chassis (q.v.) to accommodate a 25-pounder (87.6 mm) field gun.

The first examples were ready for trials by August 1941 and 100 were ordered a few months later. All production was rushed to the Middle East where it was met with some misgivings. Was the Bishop a heavily armed tank or a self-propelled gun? Eventually it was decided it was a gun and was issued to Royal Artillery batteries.

The Bishop was not a success. The gun was mounted in a large fixed superstructure which limited both traverse and elevation, the latter severely limiting maximum range. If the full range was required the crew had to build a dirt ramp and drive the vehicle up it. Ammunition stowage was limited in the cramped fighting compartment, so a trailer had to be towed. In the absence of anything better, it was deployed operationally and even took part in the early stages of the Italian campaign. Thereafter, it faded from the scene as the British began to receive sufficient numbers of the vastly superior American M7 Priest self-propelled gun.

Specification

Crew: 4
Engine power: 131 hp (97.73 kW)
Combat weight: 17,690 kg (39,006 lb)
Max speed: 24 km/h (14.90 mph)
Length: 5.53 m (18.14 ft)
Range: 145 km (90.05 miles)
Width: 2.63 m (8.63 ft)
Main gun: 25-pounder (87.6 mm)
Height: 2.825 m (9.27 ft)
Armour: 8 mm (0.32 in) to 60 mm (2.36 in)

Archer Self-propelled Gun

The Archer with its unusual rear-facing 17-pounder gun

The 17-pounder (76.2 mm) anti-tank gun began to become available from mid 1942. Produced originally as a towed weapon, the 17-pounder was soon suggested as a tank gun, giving rise to the Firefly and Challenger (q.v.). But such vehicles were far in the future when Vickers were asked to produce a tank destroyer based on the Valentine chassis.

To mount the 17-pounder gun on the relatively small Valentine, Vickers had to place it on a rearwards-facing limited traverse mounting. An open-topped armoured superstructure with a sloping front was built over the usual fighting compartment area and the long gun barrel pointed directly to the rear. The result was a fairly satisfactory tank destroyer with a compact and low silhouette. Power was provided by a GMC diesel engine. Some 800 Archers were ordered (of which 665 were delivered) and the first production model was supplied in March 1944. The first combat use occurred in North-West Europe from October 1944.

The rearwards-facing gun was not a handicap as the Archer was usually deployed in concealed positions from which it could escape if things went wrong. Its slow speed was a drawback, as it had always been for the Valentine, but the 17-pounder anti-tank gun proved to be highly effective against most German armour. What was intended to be only an interim solution until sufficient tanks with the same gun became available instead proved to be a long term success.

Specification

Crew: 4
Engine power: 165 hp (123.09 kW)
Combat weight: 16,765 kg (36,966 lb)
Max speed: 24 km/h (14.90 mph)
Length: 6.686 m (21.94 ft)
Range: 145 km (90.05 miles)
Width: 2.63 m (8.63 ft)
Main gun: 17-pounder (76.2 mm)
Height: 2.7 m (8.86 ft)
Armour: 8 mm (0.32 in) to 60 mm (2.36 in)

Churchill - Early Marks

An early Churchill mounting a Churchill ARK bridge

During the late 1930s British Army planners called for a tank capable of operating over cratered ground such as that seen on the Western Front in 1914-18. A proposal by Vauxhall Motors was accepted to become the A22, or Infantry Tank Mark IV, later known as the Churchill. This was to become one of the most important British tanks, but it was slow and its all-round-tracks were reminiscent of the Great War. Yet it was well-armoured, could traverse obstacles with ease and, as would be seen, it was highly adaptable. The first examples were completed in May 1941 but were plagued by mechanical failures which took time to rectify.

Churchills were soon in great demand. Their first action was the Dieppe fiasco of 1942 but thereafter Churchills fought on all British fronts. Their effectiveness was initially limited by the use of the 2-pounder (40 mm) gun in the cast turret - the Mark 1 also had a 3-inch (76.2 mm) howitzer in the front hull although this was soon discarded - the Mark IICS had the howitzer in the turret. The Mark III was a major redesign with a 6-pounder (57 mm) gun in a welded turret while the Mark IV reverted to the cast turret. The Mark V was armed with a 95 mm (3.75 in) howitzer while the Mark VI had a 75 mm (2.96 in) gun. The Mark VII was a major redesign (next entry).

Specification - Mark IV

Crew: 5
Engine power: 350 hp (261.1 kW)
Combat weight: 39,660 kg (87,450 lb)
Max speed: 24 km/h (14.90 mph)
Length: 7.442 m (24.42 ft)
Range: 145 km (90.05 miles)
Width: 2.743 m (9.00 ft)
Main gun: 2-pounder (40 mm)
Height: 3.25 m (10.66 ft)
Armour: 16 mm (0.63 in) to 102 mm (4.02 in)

Churchill - Later Marks

A later model of Churchill moving through a village in Italy

The Churchill Mark VII was a major redesign. The turret was enlarged, with a commander's cupola, and dust guards fitted over the tracks. Armour was increased, there was a new gearbox and numerous detail design changes. The main armament was a 75 mm (2.96 in) gun; on the Mark VIII it was a 95 mm (3.75 in) close support howitzer. The Mark IX to the Mark XI incorporated further improvements and there were numerous sub-marks such as the Mark IV NA75, a North African model created by rearming with 75 mm guns taken from wrecked Shermans.

The Churchill AVRE, a Royal Engineers 'special' armed with a mortar that launched a heavy demolition charge known as the Petard. This was only one of a host of Churchill special role tanks. To list all of them would fill many pages and one of the more important, the Churchill Crocodile, is covered in the next entry. Suffice it to say that the Churchill fulfilled just about every role an armoured vehicle could be called upon to assume. The weight and stability of the Churchill enabled it to carry all manner of equipment, from landing mats to bridges. Churchills remained in service for years after 1945.

Specification - Mark VII

Crew: 5
Engine power: 350 hp (261.1 kW)
Combat weight: 40,640 kg (89,611 lb)
Max speed: 20 km/h (12.42 mph)
Length: 7.54 m (24.74 ft)
Range: 350 km (217.35 miles)
Width: 2.74 m (8.99 ft)
Main gun: 75 mm (2.96 in)
Height: 3.35 m (10.99 ft)
Armour: 25 mm (0.98 in) to 152 mm (6 in)

Crocodile

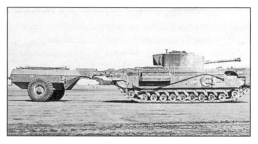

The Churchill Crocodile showing the towed fuel trailer for its bow-mounted flamethrower

The Crocodile is only one of many special purpose Churchills which could have been selected for an individual mention, but there is not enough space here to cover them all. The Crocodile was a spectacular example: a flame-thrower tank used to grim effect from June 1944 onwards.

The Crocodile towed an armoured two-wheel trailer containing 400 gallons (1818 litres) of flame fuel delivered via an articulated towing joint to a flame gun in the tank's front hull. (The flame gun took the place of the hull machine gun). Pressure was provided by compressed nitrogen and under ideal conditions it could project a flame jet up to 110 metres (120 yards) long. Maximum effective range was about 75 metres

(82 yards). The fuel was sufficient for up to 80 one-second flame bursts. If it became necessary the trailer could be rapidly jettisoned. If required, all the Crocodile equipment could be removed to allow the carrier to revert to a gun tank.

The Crocodile proved fearsomely effective, burning out the occupants of bunkers and field fortifications. Crocodiles served on until the end of the war but after July 1944 were usually reserved for set-piece assaults such as the breakthrough of the Siegfried Line. A small number of Sherman Crocodiles were produced but only four ever saw service with the US Army.

Specification

Crew: 5
Engine power: 350 hp (261.1 kW)
Combat weight: tank - 40,640 kg (89,611 kW); trailer - 6,600 kg (14,553 lb)
Max speed: 20 km/h (12.42 mph)
Length: tank - 7.54 m (24.74 ft)
Range: 350 km (217.35 miles)
Width: 2.74 m (8.99 ft)
Main gun: 75 mm (2.96 in) and flame-thrower
Height: 3.35 m (10.99 ft)
Armour: 25 mm (0.98 in) to 152 mm (6 in)

Duplex Drive System

Duplex drive during early trials on a Valentine tank

Duplex Drive equipment was developed as early as 1941, giving combat vehicles a temporary amphibious capability. It involved a collapsible screen, anchored at the base to a boat-shaped platform welded around the vehicle hull. Compressed air was routed through rubber tubes which formed pillars and raised the top level of the screen above the vehicle, forming a waterproof structure which enabled the vehicle to float.

Power in the water was provided by twin screw

propellers driven from the main engine and located under the vehicle hull rear; a rudimentary steering system was provided. Once on dry land the screen was collapsed by releasing the air in the pipes and the vehicle could then proceed as normal.

The first tank to be fitted with the Duplex Drive was a Tetrarch which was used for trials only. This was followed by numbers of Valentine infantry tanks which were supposed to be the main British Duplex Drive tank (a few did see use in Italy during 1945) but in the event it was the Sherman which employed the system during the Normandy landings. They were successful although there were casualties in the rough seas. The maximum freeboard afforded by the screen was only just under one metre. Matters were not assisted by the slow in-water speeds which were only a couple of miles an hour.

But sufficient Duplex Drive Shermans did manage to make the shore to make their presence felt and the system was considered successful enough to be utilised once more, again with Shermans, during the Rhine Crossings of March 1944.

Specification

Freeboard: 1 metre
Speed in water: Up to 5 km/h (3 mph)

Universal Carriers

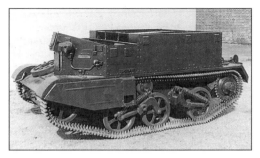

A typical Universal Carrier in its basic form

The Universal Carriers were not tanks but light armoured personnel carriers, yet they were so important to the British Army during World War II that they deserve mention. Their origins can be traced back to the Carden-Loyd tankettes of the 1920s and were gradually developed during the 1930s via a bewildering number of models and experimental types to the form originally known as the Bren Gun Carrier.

It was a completely open little vehicle with thin armour, a simple but effective suspension, and limited accommodation for the crew. The usual crew of the Universal Carrier was only three or four men, two at the front and one or two in the rear seated each side of the engine. A machine gun or anti-tank rifle was mounted over the front with a facility to mount another weapon firing through a hatch in the front plate.

This basic configuration was used for all manner of applications, from radio carrier to mortar carrier. The Wasp was a highly effective flame-thrower variant while others were used as command vehicles or anti-tank gun tractors. All manner of experimental versions appeared, the Praying Mantis being one of the oddest, a Carrier chassis coupled with an extending ladder to scale cliffs.

Universal Carriers served in Europe, during the North African campaign and in the Far East. Several thousand were shipped to Russia via the Arctic convoys and were used by the Red Army.

Specification

Crew: 3 or 4
Engine power: (typical) 85 hp (63.41 kW)
Combat weight: (max) 4500 kg (9,922 lb)
Max speed: 48 km/h (29.8 mph)
Length: 3.65 m (11.98 ft)
Range: approx 150 km (93.15 miles)
Width: 2.05 m (6.73 ft)
Main gun: variable
Height: 1.59 m (5.22 ft)
Armour: 7 mm (0.28 in) to 10 mm (0.39 in)

Medium Tank M2A1

One of the early models of the Medium Tank M2

The M2 never saw combat but still made an important contribution to the Allied victory. After neglecting military developments in Europe for many years, America began to rearm in 1939. The US Army had no modern armour, so a tank factory was built capable of manufacturing ten tanks a day. The programme was passed to the US automotive industry and within a year the Detroit Tank Arsenal was under construction

The tank design chosen was the Medium Tank M2. Intended for infantry support and armed with a 37 mm (1.46 in) anti-tank gun in a turret, it had four

machine guns in traversing sponsons located around
the hull, plus two more on the turret sides to fire
forward.

The main production model was the M2A1 with a
more powerful engine and a larger turret than the
M2. But almost as soon as the M2A1 appeared, the
Battle of France took place. It was apparent that
German tanks were more advanced than the
Americans realised: some even had 75 mm (2.96 in)
guns. Production ceased after only 94 had been
completed, in favour of the Medium Tank M3 (next
entry). But the M2A1 had already played its part by
ensuring that the Detroit Tank Arsenal was built to
produce them. Thereafter, the Detroit facility was to
churn out tanks by the thousand, altering the course
of the war.

Specification - M2A1

Crew: 6
Engine power: 400 hp (298.4 kW)
Combat weight: 21,330 kg (46,966 lb)
Max speed: 42 km/h (26.08 mph)
Length: 5.334 m (17.50 ft)
Range: 210 km (130.41 miles)
Width: 2.59 m (8.50 ft)
Main gun: 37 mm (1.46 in)
Height: 2.82 m (9.25 ft)
Armour: max 32 mm (1.26 in)

Medium Tank M3

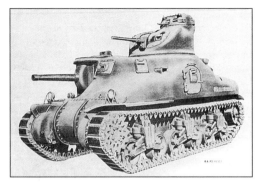

The M3 was known to the British Army as the Lee

To match the firepower of German tanks, the US Army needed a tank with a 75 mm (2.96 in) gun, but the M2 (previous entry) could not carry such a heavy weapon in its turret. The solution was to mount a 75 mm gun in a limited traverse sponson on the right-hand side of a much-modified M2 hull. A small cast turret with a 37 mm (1.46 in) gun was located on the top left.

The British ordered M3s built to their specifications as the Grant (q.v.) and bought some standard M3s (which they named the Lee). Batches of M3s were also sent to the Soviet Union. British combat experience in North Africa highlighted the disadvantages of the M3's limited traverse gun mounting, but this had always been appreciated by the US designers who intended the M3 to be a stopgap before the arrival of the M4 (next entry). At least the 75 mm gun was able to destroy German tanks invulnerable to the old 2 pounder.

By the time M3 production ceased in December 1942, 6,258 had been made, over 1,100 of them in Canada. After that time, as M3s were withdrawn they were used for a large number of trials and special purpose conversions, including the M7 105 mm (4.14 in) self-propelled howitzer (q.v.), the M12 155 mm (6.11 in) self-propelled gun (next entry), the M33 heavy artillery tractor, and a widely used armoured recovery vehicle, the M31.

Specification - M3

Crew: 6
Engine power: 340 hp (253.64 kW)
Combat weight: 27,240 kg (60,064 lb)
Max speed: 42 km/h (26.08 mph)
Length: 5.64 m (18.50 ft)
Range: 193 km (119.85 miles)
Width: 2.718 m (8.92 ft)
Main gun: 75 mm (2.96 in)
Height: 3.124 m (10.25 ft)
Armour: 12.7 mm (0.50 in) to 57 mm (2.25 in)

Gun Motor Carriage M12

The Gun Motor Carriage M12 carrying 155 mm gun

In mid-1941 US Army Artillery staff requested a 155 mm (6.11 in) gun which could be mounted on the M3 medium tank chassis (q.v.). A design using a long-barrelled 155 mm M1918 gun was obtained and despite some official opposition, the artillery staff persisted and an order was placed for 100 examples of the Gun Motor Carriage M12.

The conversion of the M3 was straightforward: the engine was moved forward, leaving the rear of the hull free for to accomodate the gun mounting and leaving space into which the breech could be lowered. The gun mounting was open, but since the 155 mm gun was a long range weapon, the crew would not (in theory) be exposed to enemy fire.

The US Army continued to ignore the need for the M12, so the 100 built were initially used only for training within the United States. However, by late 1943, with the invasion of Europe looming, it was decided that there was a need for the M12 after all. A total of 75 M12s were rebuilt prior to being sent to North-West Europe. The Cargo Carrier M30 was identical to the M12 apart from the omission of the gun; it was used as an ammunition carrier for M12 batteries on a one-to-one basis since the M12 could carry only 10 rounds ready to fire. Most of the gun crew also travelled on the M30.

Specification

Crew: 6
Engine power: 400 hp (298.4 kW)
Combat weight: 26,650 kg (58,763 lb)
Max speed: 32 km/h (19.87 mph)
Length: 6.7 m (21.98 ft)
Range: 300 km (186.3 miles)
Width: 3.86 m (12.66 ft)
Main gun: 155 mm (6.11 in)
Height: 2.69 m (8.83 ft)
Armour: 9.4 mm (0.37 in) to 25.4 mm (1.00 in)

Medium Tank M4

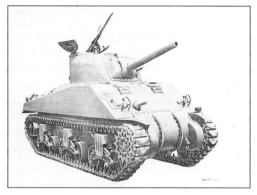

M4A – An early production Medium Tank M4 with standard 75 mm gun

As soon as design of the M3 was completed, thoughts turned to producing a tank with a turret-mounted 75 mm (2.96 in) gun. A prototype, the Medium Tank T6, appeared in September 1941. Entering production as the M4 at the incredible rate of 1,000 (later 2,000) a month from 11 plants, by 1946 over 40,000 had been built - a decisive contribution to Allied victory.

Based on the hull of the M3, the M4 had the same engine and suspension but M3's side doors were omitted. Engine supply was a problem: the intended Wright Continental radial was also needed for aircraft. Several others were used, the Chrysler Multibank engine of the M4A4 (actually five automobile engines driving a common main shaft) was so big that the hull had to lengthened.

M4s were immediately supplied to the British, who named it Sherman; this became the most numerically important of all British tanks. Entire Soviet tank corps were also equipped with M4s. From early 1944 M4s were armed with more powerful 76 mm (2.99 in) guns or even 105 mm (4.14 in) howitzers. The M4 was adapted to fulfil many roles and carry many types of special equipment, ranging from flame-throwers and mine rollers to dozer blades and rocket launchers.

Specification - M4A1

Crew: 5
Combat weight: 30,160 kg (66,502 lb)
Length: 5.84 m (19.16 ft)
Width: 2.68 m (8.79 ft)
Height: 2.97 m (9.74 ft)
Engine power: 400 hp (298.4 kW)
Max speed: 38 km/h (23.60 mph)
Range: 160 km (99.36 miles)
Main gun: 75 mm (2.96 in)
Armour: 25 mm (0.98 in) to 51 mm (2.01 in)

Medium Tank M4 - Late Production

M4B – A late production Medium Tank M4 complete with uprated suspension, long 76 mm gun and dust shields

Standardised in February 1942, the Medium Tank M4A3 was typical of the later M4s. It had the 76 mm (2.99 in) (actually 76.2 mm (3 in)) main gun as standard (although the 75 mm (2.96 in) gun continued to be used on some vehicles), more armour, and a Ford V-8 gasoline unit specifically designed for tanks. The suspension springing was improved, with the track support rollers located behind the suspension units.

It was adapted for many roles, including close support with a 105 mm (4.14 in) howitzer. The

M4A3s were the most important tanks of all World War II combat tanks. High, and with less armour and firepower than late war German tanks, their gasoline engines were dangerously inflammable. But they were less prone to breakdown than the theoretically superior German vehicles. Crews added track links, logs and sandbags to increase protection.

While the M4A3 had the Ford V-8 engine the M4A4 had the Chrysler Multibank. The M4A5 was the designation applied to the Canadian Ram (q.v.) while the Caterpillar-engined M4A6 was produced in small numbers only (75). Well over 5,800 early models were reworked from mid to late 1944 onwards to bring them up to at least M4A3 standard. Various models of M4 were used to produce the widely used M32 armoured recovery vehicle.

Specification - M4A3

Crew: 5
Combat weight: 32,285 kg (71,188 lb)
Length: overall, 7.518 m (24.67 ft)
Width: 2.68 m (8.79 ft)
Height: 3.246 m (10.65 ft)
Engine power: 500 hp (373 kW)
Max speed: 47 km/h (29.19 mph)
Range: 160 km (99.36 miles)
Main gun: 76.2 mm
Armour: 38 mm (1.50 in) to 63.5 mm (2.50 in)

Gun Motor Carriage M10

Gun Motor Carriage M10 with the 3-inch gun and turret pointing to the rear

In the US Army, tank destroyers were seen as a separate branch of armour and had their own Tank Destroyer Command. It was decided to convert a M4A2 hull powered by twin GM diesels and combine it with an open-topped welded turret mounting a converted 3-inch (76.2 mm) anti-aircraft gun. The top of the M4A2 hull was flattened to reduce height slightly and armour was reduced to save weight and improve its mobility. After a while the balance of the turret was improved by adding counterweights to the rear. There was no hull machine gun although a 0.50 (12.7 mm) machine gun could be pintle-mounted on the turret side.

The result was the Gun Motor Carriage M10, standardised in June 1942. About 7,000 were built, some 2,600 of them based on the M4A3 chassis and known as the M10A1. The latter batch was never sent overseas and were used either for training or were converted to M35 prime movers for heavy artillery. Production ceased in December 1943 as attentions were switched to the M36 (next entry).

The original M10s saw hard action in the Normandy campaign. M10s were part of the Lease-Lend arrangements with the United kingdom; a further 52 were sent to the Soviet Union. In British service the M10 was known as the Wolverine but the 3-inch gun was replaced by the same 17-pounder anti-tank gun used in the Sherman Firefly and Archer tank destroyer.

Specification

Crew: 5
Combat weight: 29,938 kg (66,013 lb)
Length: 5.97 m (19.59 ft)
Width: 3.05 m (10.00 ft)
Height: 2.477 m (8.13 ft)
Engine power: 450 hp (335.7 kW)
Max speed: 48 km/h (29.8 mph)
Range: 320 km (198.72 miles)
Main gun: 3-inch (76.2 mm)
Armour: 12 mm (0.47 in) to 37 mm (1.46 in)

Gun Motor Carriage M36

Gun Motor Carriage M36, a tank destroyer armed with a 90 mm gun able to knock out most German tanks

Effective as the M10 tank destroyer (previous entry) was, it was felt that a gun with better long-range performance than the 3-inch (76.2 mm) would be required. There was a 90 mm (3.55 in) anti-aircraft gun available, but a new turret had to be designed to mount this larger weapon. The result was the Gun Motor Carriage M36 standardised in June 1944.

An order for 500 M36s was soon placed. The initial production run used the Ford V-8 engined M10A1 chassis although there was an expedient model, the

M36B2, which involved placing the M36 turret directly onto unconverted M4A3 medium tanks; 187 of these conversions were made. About 500 M10A1s were converted to M36 standard in order to get M36s into service as fast as possible. The M36 proved itself to be one of the most powerful of the American tank destroyers, much better than the M10 which was withdrawn and replaced by M36s.

In action by late 1944, the M36's high-velocity 90 mm gun and its specially developed armour-penetrating ammunition proved able to destroy Panthers and Tigers at long range, ending the era of the German 8.8 cm (3.47 in) anti-tank gun's battlefield supremacy. Some M36 units were able to record very impressive scores with few losses to themselves.

Specification

Crew: 5
Combat weight: 28,120 kg (62,004 lb)
Length: 6.15 m (20.18 ft)
Width: 3.05 m (10.00 ft)
Height: 2.72 m (8.92 ft)
Engine power: 450 hp (335.7 kW)
Max speed: 48 km/h (29.8 mph)
Range: 240 km (149.04 miles)
Main gun: 90 mm (3.55 in)
Armour: 12 mm (0.47 in) to 50 mm (1.97 in)

Gun Motor Carriage M18 Hellcat

The Hellcat was one of the fastest tank destroyers

In December 1941 the US Army's Tank Destroyer Command proposed a 37 mm (1.46 in) gun-armed vehicle with a torsion bar suspension and an open turret. This was later revised to allow a 75 mm (2.96 in) gun to be carried and then a 76 mm (2.99 in) gun, to keep in step with successive M4 medium tank series (q.v.) armament changes. The final form of a series of experimental vehicles was designated T70 and in July 1943 the T70 became the Gun Motor Carriage M18 and went into production; 1,000 were initially ordered although by October 1944, when production ceased, 2,507 had been built.

The M18 was one of the fastest armoured fighting vehicles of World War II, having a maximum road speed of 80 km/h (49.68 mph). This was mainly due to a combination of light weight, a sturdy Christie-based suspension, and a 400 hp (298.4 kW) engine, all combining to provide a high power-to-weight ratio. The light weight was partly due to the light armour provided, for it was intended that the M18 would utilise its speed and mobility for 'shoot and scoot' tactics; the light armour was partially offset by the arrangement of the sloping armour plates to provide extra protection.

Crews named the M18 the Hellcat, although the name was unofficial. It was later rated as the best all-round tank destroyer of World War II.

Specification
Crew: 5
Combat weight: 18,140 kg (39,998 lb)
Length: 6.65 m (21.82 ft)
Width: 2.97 m (9.74 ft)
Height: 2.565 m (8.42 ft)
Engine power: 400 hp (298.4 kW)
Max speed: 80 km/h (49.68 mph)
Range: 240 km (149.04 miles)
Main gun: 76 mm (2.99 in)
Armour: 7 mm (0.28 in) to 12 mm (0.47 in)

Howitzer Motor Carriage M7

A factory-fresh M7, known to the British as the Priest

Almost as soon as the Medium Tank M3 was off the drawing board, plans were being made to convert its chassis to accommodate a 105 mm (4.14 in) howitzer for issue to armoured artillery units supporting tank operations. The upper hull was drastically modified to mount the 105 mm howitzer in an off-centre position. A large drum-like cupola was added to the right-hand superstructure to allow a 0.50 in (12.7 mm) machine gun to be placed on a ring mounting for air and local defence. Trials vehicles were known as the T32 and by February 1942 it was standardised as the Howitzer Motor Carriage M7.

Almost as soon as the first M7s were ready they were

seen by a visiting British mission which immediately requested 2,500, soon to be followed by a request for a further 3,000. Although these requests were never fully met as the requirements of the US Army had priority, the M7 became an important British artillery piece. With the British it was known as the Priest, due to the pulpit-like side cupola. The first went into action at El Alamein. Priests were used in Italy and went on to take part in the Normandy landings when they were provided with deep wading equipment.

When the M3 tank was replaced by the M4, production of M7 hulls also switched to the new chassis; these models were known as the M7B1. Although M7s were still in US Army service in 1945, by then they had been phased out from British service in favour of the Sexton.

Specification

Crew: 7
Combat weight: 22,695 kg (50,042 lb)
Length: 6.02 m (19.75 ft)
Width: 2.87 m (9.42 ft)
Height: 2.54 m (8.33 ft)
Engine power: 450 hp (335.7 kW)
Max speed: 40 km/h (24.84 mph)
Range: 200 km (124.2 miles)
Main gun: 105 mm (4.14 in)
Armour: 12 mm (0.47 in) to 62 mm (2.44 in)

Howitzer Motor Carriage M8

Howitzer Motor Carriage M8, showing the stubby 75 mm howitzer in the front of the open turret

The Howitzer Motor Carriage M8 was developed following a request for a light close support vehicle to operate within tank battalions. The Light Tank M5 fitted with a 75 mm (2.96 in) Pack Howitzer in a larger, open-topped fully-traversable turret with the 75 mm howitzer in a mantlet which allowed a barrel elevation angle of up to +40 degrees.

This was adapted as the Howitzer Mounting Carriage M8 and went into mass production under Cadillac supervision - 1,778 were produced between

September 1942 and January 1944. The M5 hull was employed virtually unchanged. The M8A1 introduced the modifications made for the Light Tank M5A1 (q.v.).

The M8 proved to be a highly serviceable fire support vehicle. Virtually all were used only by the US Army and served mainly in North-West Europe from June 1944 onwards. The M8's one drawback for the self-propelled howitzer role was that internal ammunition stowage space was limited, so M8s were often observed towing ammunition trailers carrying extra rounds. M4 mediums armed with 105 mm (4.14 in) howitzers became available from late 1944 onwards and gradually replaced the M8, but by 1945 the M8 was still in front line use. A few still serve today with some Third World armed forces.

Specification

Crew: 4
Combat weight: 16,330 kg (36,007 lb)
Length: 4.84 m (15.88 ft)
Width: 2.242 m (7.36 ft)
Height: 2.23 m (7.32 ft)
Engine power: 220 hp (164.12 kW)
Max speed: 60 km/h (37.26 mph)
Range: 160 km (99.36 miles)
Main gun: 75 mm (2.96 in)
Armour: 12 mm (0.47 in) to 67 mm (2.64 in)

Light Tank M2

The M2A2 was only armed with twin machine guns

The US Army was not officially supposed to have light tanks prior to 1939. Tanks were intended to be infantry support vehicles. So the US Cavalry developed their own light tanks under the disguised designation of 'Combat Car' M1 and M2. Following a change of philosophy during the late 1930s these were developed further into the Light Tank M2 series. Although termed light tanks, the M2s were much larger and better-armed than many similar vehicles of their time. The main production model, the M2A4, had a 37 mm (1.46 in) anti-tank gun and a coaxial machine gun (early models had only two turret machine guns) with two more machine guns in small side sponsons and another in the front hull. Overall, the M2 series were simple, lightly armoured tanks, little more than reconnaissance vehicles.

Production of the Light Tank M2 series was limited (365 were manufactured by March 1941), even though it was considered that they were simple enough to be mass-produced in haste by heavy industry from October 1939 onwards. Thus by the time war commenced with Japan and Germany in 1941 the M2 was in service with the US Army. A few saw some service in the Philippines in 1942 but the main production run was used for training American and British tank crews before they went into battle in rather more modern vehicles.

Specification - M2A4

Crew: 4
Combat weight: 10,430 kg (22,998 lb)
Length: 4.42 m (14.50 ft)
Width: 2.47 m (8.10 ft)
Height: 2.49 m (8.17 ft)
Engine power: 250 hp (186.5 kW)
Max speed: 48 km/h (29.8 mph)
Range: 200 km (124.2 miles)
Main gun: 37 mm (1.46 in)
Armour: 6 mm (0.24 in) to 25 mm (0.98 in)

Light Tank M3

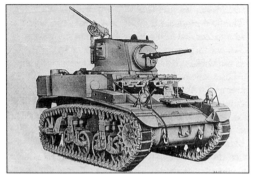

Light Tank M3, known to the British Army as the Stuart

US observers in France in 1940 noted that the M2 series would have to be improved to remain a viable combat vehicle. Thus the M2A4 was provided with better armour, the suspension was strengthened and the fixed rear trailer wheel was lowered to increase the length of track on the ground. The M3 model became the most important of the American light tanks as the eventual production total, from March 1941 until August 1942, was 5,811. They served almost everywhere, from the Eastern Front to the Pacific, and were adopted by the British Army where they were officially known as Stuarts - the alternative name Honey was unofficial.

The final production model was the M3A3 which had a redesigned hull outline - 3,427 of them were built, not forgetting the 4,621 for the M3A1 (there was no M3A2). The M3 series was adapted for many purposes, one of the simplest being the removal of the turret and replacing it with a welded superstructure for use as a command vehicle. In the Pacific the US Marines used old M3s with flame-throwers in place of the 37 mm (1.46 in) gun.

Many M3 series light tanks were still in service when the war ended in 1945. Widely exported after the war, they continued to serve all over the world. Many received extensive modification to keep them running, the Brazilian army re-building them into effectively new tanks as late as the 1980s.

Specification - M3A1

Crew: 4
Combat weight: 12,925 kg (28,499 lb)
Length: 4.54 m (14.90 ft)
Width: 2.235 m (7.33 ft)
Height: 2.3 m (7.55 ft)
Engine power: 250 hp (186.5 kW)
Max speed: 60 km/h (37.26 mph)
Range: 110 km (68.31 miles)
Main gun: 37 mm (1.46 in)
Armour: 10 mm (0.39 in) to 51 mm (2.01 in)

Light Tank M5

Light Tank M5 fitted with wading gear for amphibious operations

Although the Light Tank M3 series were produced by the thousand, production was troubled by shortages of all manner of items, especially engines. One solution was to install twin Cadillac V-8s coupled to a Cadillac Hydra-matic automatic transmission. The new drive train made the vehicle easy to drive, and the new version was standardised as the Light Tank M5 in February 1942. (The M4 designation was not used to avoid confusion with the Medium Tank M4.)

The M5 went into production on Cadillac automobile production lines - Cadillac were the prime contractor. The M5A1, introduced from early 1943 onwards, was essentially the same vehicle as the original M5 but with a larger turret to provide space for radios. Most M5s were used by American forces but a few served with the British Army from late 1944.

Attempts were made to convert turretless M5s into mortar carriers but none were successful - the 75 mm (2.96 in) M8 self-propelled howitzer was a major M5 variant (next entry). M5 command vehicles were produced with fixed superstructures in place of the turret while the T8 was a 1944 reconnaissance vehicle created by removing the M5 turret and fitting a 0.50 in (12.7mm) machine gun mounting ring in its place. Although never officially standardised by the US Army, the T8 and similar T8E1 were employed in some numbers after 1945.

Specification - M5A1

Crew: 4
Combat weight: 15,380 kg (33,912 lb)
Length: 4.84 m (15.88 ft)
Width: 2.242 m (7.36 ft)
Height: 2.23 m (7.32 ft)
Engine power: 220 hp (164.12 kW)
Max speed: 60 km/h (37.26 mph)
Range: 160 km (99.36 miles)
Main gun: 37 mm (1.46 in)
Armour: 12 mm (0.47 in) to 67 mm (2.64 in)

Light Tank M22

Light Tank M22, one of the few armoured vehicles designed specifically for the airborne role

As early as February 1941 American tank staff had recognised the need for an airborne tank. The T9E1, a design from Marmon-Herrington was adopted and ordered into production in early 1943. By February 1944, 830 had been produced. In September 1944 it became the Light Tank, (Airborne) M22. By then it had become apparent that the M22 was never likely to be used in its intended role since there were no aircraft or gliders in US service large enough to carry it.

The M22 was therefore handed over to the British airborne forces who did have a glider, the Hamilcar, with a fuselage large enough to carry such a vehicle. With the British, the M22 was known as the Locust and a handful were used during the March 1945 Rhine crossings but they were of limited combat value and were withdrawn soon after.

The main problem for the M22 was that its necessarily low weight severely limited the armour which could be provided. Although the main armament was a 37 mm (1.46 in) anti-tank gun the M22 was no match for any of the German tanks of 1945 and would have proved a deathtrap in any combat encounter. This problem was to dog airborne tanks until the development of anti-tank missiles. Airborne tanks of the immediate post-war era, such as the Russian ASU series would have been hopelessly outmatched by any enemy tank they encountered.

Specification
Crew: 3
Combat weight: 7,445 kg (16,416 lb)
Length: 3.937 m (12.92 ft)
Width: 2.16 m (7.09 ft)
Height: 1.854 m (6.08 ft)
Engine power: 162 hp (120.85 kW)
Max speed: 64 km/h (39.74 mph)
Range: 210 km (130.41 miles)
Main gun: 37 mm (1.46 in)
Armour: 9 mm (0.35 in) to 25 mm (0.98 in)

Light Tank M24 Chaffee

Light Tank M24, armed with the remarkable 75 mm M6 gun continued to serve in many armies after 1945

By 1942 it was becoming apparent that tanks had to have at least a 75 mm (2.96 in) gun. Attempts to fit such weapons into M3/M5 series light tanks were unsuccessful due to their narrow hulls. A new design emerged, the T24, which used many automotive features of the M5, allied to a new suspension. To avoid the vehicle becoming overweight the armour was thin, but well sloped, and a new turret housing a 75 mm M6 gun. This was a remarkable weapon, originally an anti-aircraft gun derived from a 75 mm field gun dating from 1897! Its light weight and compact dimensions made the 75 mm M6 suitable for mounting in a light tank turret.

The T24, standardised in late 1943 as the Light Tank M24 Chaffee, was ordered into mass production with orders eventually reaching 5,000. The first M24s did not reach Europe until late 1944. They proved highly effective and, thanks to their well-tried automotive systems, highly reliable. Variants included an air defence version with twin 40 mm (1.58 in) Bofors guns in an open ring mounting, the M37 105 mm (4.14 in) howitzer carriage and the M41 155 mm (6.11 in) howitzer carriage.

After 1945 the M24 Chaffee served with many US allies. The French army used them in Indo-China, one squadron fighting with great distinction at the battle of Dien Bien Phu.

Specification

Crew: 5
Combat weight: 18,370 kg (40,505 lb)
Length: overall, 5.49 m (18.01 ft)
Width: 2.946 m (9.67 ft)
Height: 2.477 m (8.13 ft)
Engine power: 220 hp (164.12 kW)
Max speed: 56 km/h (34.78 mph)
Range: 160 km (99.36 miles)
Main gun: 75 mm (2.96 in)
Armour: 9 mm (0.35 in) to 25 mm (0.98 in)

Heavy Tank M6

One of the heaviest tanks yet designed in 1942, the M6 was destined never to leave the USA

The Heavy Tank M6 never saw action but was important since it provided the design information for further development leading to greater things. Conceived as early as May 1940 the M6 was ready for production in April 1942. A large and powerful tank which would no doubt have been a match for the Panther, or even the Tiger, its main gun was a converted 3-inch (76.2 mm) anti-aircraft gun and there was even a coaxial 37 mm (1.46 in) anti-tank gun.

Unfortunately, at that time the USA was frantically reconsidering manufacturing priorities, and the emphasis was placed on medium tank production.

Shipping the large M6 overseas would have consumed scarce shipping space needed for other weapons.

With this in mind the M6 was re-evaluated and found wanting. The internal layout was deemed unsatisfactory, the complex turret was too small for further development, the 3-inch gun was (by 1942) thought to be obsolescent, and the engine/drive train was still not fully developed and too unreliable for prolonged combat use.

Thus M6 production was limited from the planned 5,000 to only 40. Yet those 40 were utilised for the many trials and experiments which led to the highly successful M26 heavy tank series (next entry). That was perhaps the M6's greatest contribution to American tank development during World War II.

Specification

Crew: 6
Combat weight: 57,380 kg (126,522 lb)
Length: 8.433m (27.67 ft)
Width: 3.11m (10.20 ft)
Height: 3.226m (10.58 ft)
Engine power: 800 hp (596.8 kW)
Max speed: 35 km/h (21.74 mph)
Range: 160 km (99.36 miles)
Main gun: 1 x 3-inch (76.2 mm); 1 x 37 mm (1.46 in)
Armour: 25 mm (0.98 in) to 100 mm (3.94 in)

Heavy Tank M26 Pershing

Side view of the Heavy Tank M26, displaying to good effect its long-barrelled 90 mm gun

The Heavy Tank M26, named the Pershing, started out as the T26 medium tank, one of a series of trial vehicles which followed the Medium Tank M4 (q.v.). The T26 was very different, featuring the use of a Torque-matic drive resulting from trials carried out on the M6 (previous entry). A need to improve the armament increased weight so much that the project became a heavy tank. The result was the M26.

The first M26s reached Europe in February 1945 and were soon in great demand, for at last the US Army had a tank which could meet the German Tiger and Panther on equal terms, and its mobility was greater than either of those tanks. After VE-Day M26s saw service in the Pacific and were employed during the invasion of Okinawa. No Japanese tank was able to take on a Pershing (most were dug in as part of the island's static defences) and only the heaviest Japanese anti-tank guns had any chance of penetrating its armour.

The M26 went on to a long post-war period of development which was ultimately to lead to the M48 and M60 series of tanks which are still in service with many US allies today. With its a long-barrelled 90 mm (3.55 in) main gun, heavy armour and a powerful Ford V-8 petrol engine, the M26 was the best US tank of World War II. The torsion bar suspension was sturdy and capable of further development while many sub-systems, such as the stabilised gun fire control system, were considerably in advance of foreign combat vehicles.

Specification

Crew: 5
Combat weight: 41,730 kg (92,014 lb)
Length: overall, 8.788 m (28.83 ft)
Width: 3.505 m (11.50 ft)
Height: 2.77 m (9.09 ft)
Engine power: 500 hp (373 kW)
Max speed: 48 km/h (29.8 mph)
Range: 148 km (91.91 miles)
Main gun: 90 mm (3.55 in)
Armour: 51 mm (2.01 in) to 101.6 mm (4.00 in)

Landing Vehicles Tracked (Armoured)

An LVT(A) fitted with a 75 mm howitzer

The Landing Vehicles Tracked (Armoured), or LVT(A)s, were not tanks but armed amphibious armoured personnel carriers. The first of them was adapted from the Alligator, a tracked rescue vehicle for use in swampy areas. The US Marine Corps adapted the concept and first used open mild steel carriers during the Tarawa landings of November 1943.

The bloody battle for Tarawa highlighted the need for armoured protection. Despite the ferocious preliminary bombardment, enough Japanese anti-tank guns remained in action to engage the landing craft. The first armoured version, the LVT(A)1, was typical of what was to follow. In place of the usual open cargo area between the two sets of tracks, an armoured deck supported a small turret mounting a 37 mm (1.46 in) gun. It looked high and awkward but proved ideal for providing fire support amphibious operations.

The LVT(A)2 was an open supply carrier. The LVT(A)3 and LVT(A)4 had a turret mounting a short 75 mm (2.96 in) howitzer taken from the Gun Motor Carriage M8 (q.v.) and the machine gun positions were eliminated. Extra flotation bags had to be added for these models to compensate for the extra weight as their freeboard was reduced to a dangerous level. The later LVT(A)5 was essentially a LVT(A)4 with powered controls for the turret. Some LVT(A)1s were fitted with flame-throwers.

Specification - LVT(A)1

Crew: 4
Combat weight: 14,880 kg (32,810 lb)
Length: 7.95 m (26.08 ft)
Width: 3.25 m (10.66 ft)
Height: 3.07 m (10.07 ft)
Engine power: 250 hp (186.5 kW)
Max speed: 40 km/h (24.84 mph)
Range: 200 km (124.2 miles)
Main gun: 37 mm (1.46 in)
Armour: 6 mm (0.24 in) to 13 mm (0.51 in)

Armoured Tactics 1939-45

We have grown so used to the general term Blitz, or Blitzkrieg, that it has come to mean many things to many people. To Poland in 1939 and France in 1940 the term meant 'Lightning War' and defeat. To the British it meant the air raids of 1940 and 1941. The term also came to be liberally applied to all the German tank tactics of World War II, but the latter application is incorrect for although the German Army introduced the Blitzkrieg concept it did not last long: by 1942 it was outmoded.

For the origins of the armoured warfare tactics of 1939 to 1945 one has to look back to a time before

The giant Char 2C built on French experience in 1917-18

All 6 Char 2s were destroyed in 1940 by Ju-87 Stukas

the tank appeared. Some pundits refer to the chariot warfare of the Ancient World and others to the massed cavalry charges of the Mongols. In more practical terms one can look to the Sturmtruppen tactics developed by the German Army between 1916 and 1918. The Sturmtruppen (Stormtroops) were elite infantry units. By selecting experienced, fit and well-motivated soldiers and training them to maintain the momentum of an attack, bypassing heavily defended positions, a relatively few Sturmtruppen, could penetrate deep into the enemy's rear areas, disrupting communications and supply routes and attacking command posts, causing havoc and

237

Many 1930s designs placed the main gun in the hull

confusion behind the front lines.

The key to these tactics was mobility. Once the Sturmtruppen stopped to confront any particular position they lost their momentum and the attack faltered. The early use of these special units during the battle of Verdun demonstrated the effectiveness and they soon spread throughout the German Army, culminating in the massive German successes of the Spring 'Peace' Offensive of 1918. By that time the Sturmtruppen concept had been expanded to encompass whole divisions.

The German 1918 offensives were eventually beaten as much by the exhaustion of the Sturmtruppen as the ability of the Allies to plug the gaps torn in their lines. After the final German attack had been beaten, the Allied armies passed to the offensive themselves. The Allied tanks came into their own in the summer of 1918 as open warfare succeeded nearly four years of stalemate in the trenches.

The tank tactics of 1939-45 owed little to those of 1918 when the tank was only seen as an infantry support weapon. The mechanical limitations and of the early tanks allowed them to be little else. Tactics were limited to moving forward alongside infantry and firing on enemy positions when targets presented themselves. Tank rarely met tank, as the Germans only built a handful.

Massed tank armies

After World War I the armies disbanded, but a few staff officers, especially in Germany, carried out an extensive analysis of the tactical, equipment and other demands which would be needed in the next conflict. In the United Kingdom military philosophers such as Liddell Hart, Martell and Fuller expounded the advantages of tank armies where massed tanks could batter their way deep into enemy territory and create havoc. General Hans von Seekt, commander of the

German army in the early 1920s, took the message to heart. He had observed the successes of the Sturmtruppen and took their concept of a permanent state of advancing mobility as a starting point. Von Seekt realised that the Sturmtruppen were only halted when their physical capabilities had been exhausted - they were infantry after all and had to move on their own legs at a pace which carried them forward only short distances. By combining their ever-advancing tactics with the potential of the armoured vehicle von Seekt was able to forecast the basic outline of the general tactics to be adopted by the Panzer divisions of over a decade later. The French, British and American staffs virtually ignored the Sturmtruppen tactics.

German Soviet co-operation

But von Seekt had determined only a basic outline. Details had to be established before his vision of the future could become reality. German officers were therefore sent far and wide to observe and learn what others were doing with tanks. Contacts were soon made with the Soviet Union who had suffered from the new German tactics prior to their 1917 revolution and were anxious to put the new Red Army on a war footing after their Civil War of 1919-21. German and Soviet officers not only came to study together

The Red Army ordered thousands of amphibious tanks

but they turned their combined attentions to the types of equipment the future would require.

During this exploratory period the emphasis was on mobility. Tank formations had to keep moving, but tanks alone could not provide the space for manoeuvre that would be required. The 'all tank' formation approach expounded by Liddell Hart and Fuller was therefore adapted towards a more balanced force which would include infantry travelling on trucks or light armoured vehicles along with combat engineers and artillery to clear a path for the tanks. If artillery was not available, aircraft could take their place, as Junkers Ju-87 Stuka dive bombers were to do in 1939-40.

Cleaning the 50 mm gun of a Panzer III in Russia

The Germans were not alone in their thinking. In France, United Kingdom and the Soviet Union some similar developments were forecast but were not put into practice. In the Soviet Union progress towards a balanced armoured force was disrupted by Stalin's dreadful purges in which most of the army's tank experts were executed. In France the Maginot Line came to be regarded as the main defence for the nation and funding for armoured developments suffered. In the United Kingdom funds were lacking throughout the 1920s and early 1930s, but by the late 1930s things had changed appreciably. The British Expeditionary Force sent to France in 1940 was the

British cruiser tanks abandoned in Greece during 1941

only fully mechanised force to take the field - all other armies, including the German, still retained horses for many purposes, from supply to cavalry reconnaissance and towing artillery.

By the late 1930s the German Army was receiving their first combat-worthy tanks. . Reconnaissance troops, initially on motor cycles but later including light tanks or armoured cars, were added to the new Panzer divisions to probe for weak spots in enemy defences. Once discovered, the tanks would concentrate at those points, artillery and aircraft would hammer selected targets and the tanks would then roll forward with combat engineers clearing the way.

243

It all sounds simplistic now but during the late 1930s such tactics were revolutionary. They were also devastating against Poland in 1939. The sheer momentum of the Panzer formations enabled them to punch their way through the Polish front lines. Once through they spread out deep into the rear, bypassing any centres of opposition which were left for follow-up forces to reduce, and having their advances cleared by the Luftwaffe who also carried fuel and other supplies deep into the enemy rear areas to keep the Panzer columns rolling. A campaign which would normally have lasted months or longer was over within a few weeks. The term Blitzkrieg became common currency.

France and North Africa

Yet in May 1940 the French Army could only consider distributing its precious tanks along all lines to 'stiffen up' the defences. It soon became apparent that the only way to stop the German tank formations was with other tank formations, as the British were able to do at Arras. Many French tanks never fired a shot for the Panzer advances simply bypassed them, and thousands were captured by the Germans when France surrendered.

Similar Panzer tactics were used against the British in North Africa, although until sufficient tanks and

Italian tanks were badly outclassed in North Africa

artillery could be mustered by the British to counter them. In North Africa the numbers involved were small compared to what was to be the last Blitzkrieg campaign of all, Operation Barbarossa, the invasion of the Soviet Union in 1941.

The Russian Front

The German invasion of Russia was a spectacular success at first. The Russian forces were taken by surprise. Nearly all the capable officers who might have been able to counter the German advances had been removed during the purges and those remaining

245

The Russian T-34 proved almost invulnerable to the standard German 37 mm anti-tank gun (in foreground)

The Pz IV was progressively improved during the war

were reluctant to take independent action. Eventually, the German advances stuttered to a halt deep within Russia, more from lack of supplies and exhaustion than Red Army countermoves.

Although it was not apparent at that time, the 1941 advances were the last major Blitzkrieg successes. Although the Germans were to repeat their tactics, sometimes on a massive scale, such as during the advances into Ukraine during 1942, the tide was turning. The Red Army reorganised its armoured forces and re-equipped with newer and more powerful

The magnificent lines of the Panther, built by the Germans to counter the Russian T-34 series

tanks such as the formidable T-34 and KV-1. German armoured tactics therefore entered a new phase. With the Panzers also challenged by short range anti-tank weapons and tank-killer squads, infantry frequently had to support tank operations rather than the other way around. Tanks on all sides started to grow extra stand-off armour and tank commanders learned to proceed into built-up or wooded areas only with extreme caution.

The battle of Stalingrad

The main turning point for the German Army came at Stalingrad in late 1942 but the Panzers were little involved in the prolonged urban warfare of that campaign. Panzer formations went on the defensive to plug the gaps created by the Red Army breakthrough that trapped Paulus's 6th Army but were unable to free their trapped comrades. It was Germany's first major defeat, coming as it did hard on the heels of the Second El Alamein battle in Egypt. By the following Summer the German Army was ready once more for the offensive, this time against the Kursk Salient. There was little opportunity for Blitzkrieg tactics for the Battle of Kursk more closely resembled the set-piece battles of the Great War, only with more modern weapons. The battle involved the largest clash of armour the world has ever seen but it was all head-

German assault guns on the Russian front, 1943

on, close-quarter combat in which refined tactics had no place. From the battle the Red armour emerged victorious. The Panzers were thereafter to remain on the defensive as they were gradually pushed back to Berlin.

From time to time, for example in the Ardennes Offensive of December 1944, the Panzers were able to obtain a local advantage, but the growth of Allied air power and the numerical weakness of the Panzer formations made such successes increasingly rare. By 1944 the aircraft, which had once assisted the Panzers in their headlong advances, was a major anti-armour

251

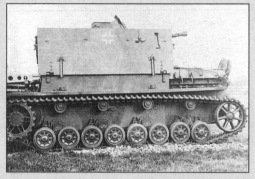

A Panzer IV converted to carry anti-aircraft guns

weapon. Strike aircraft armed with rockets or large-calibre guns were increasingly used against German tanks and other armoured vehicles making movement dangerous whenever Allied air supremacy prevailed. Another powerful anti-armour weapon emerged as concentrated artillery fire which was often able to break up tank attacks before they started.

From 1943 onwards the tank was increasingly used in close association with infantry, thereby virtually returning to the situation involved at the end of the Great War and one around which modern armoured warfare tactics still revolve. Of course there were more

A German Stug III assault gun captured by the Russians

large-scale armoured formation movements to come. The break-out from the Normandy beachhead by Patton's tank divisions was one such occasion but thereafter the war continued along the lines of set-piece battles with their artillery overtures and infantry attacks supported by tanks and assault guns.

By 1945 the face of the battlefield had changed. The days of sweeping advances were over but the tank still reigned supreme, even if it had been joined by numerous other types of armoured vehicle. The infantry, combat engineers, signallers and artillery now all proceeded about the battlefield behind

253

LVT(A)s doubled as personnel carriers/assault guns

armour, for in order to survive on a battlefield dominated by firepower produced by modern artillery, small arms and strike aircraft it had become impossible to operate in any other way.

Well before the war ended the Allies were practising the basics of inter-arm team operations in a form that survives in most modern armies to this day. The tank is still a linchpin in such team tactics but it cannot operate in isolation. The armoured personnel carrier has emerged as a major factor in tactics and is far removed in design and concept terms from the little Universal Carrier and Chenilette Lorraine of 1940.

The M4 Sherman soldiered on long after World War II

COLLINS GEM
BABIES' names

COLLINS GEM
BEER

COLLINS GEM
BIRDS

COLLINS GEM
CALORIE Counter

COLLINS GEM
FACT FILE

COLLINS GEM
FENG SHUI

COLLINS GEM
FLAGS

COLLINS GEM
Healthy **EATING**

COLLINS GEM
QUOTATIONS

COLLINS GEM
SAS Self-Defence

COLLINS GEM
SAS Survival Guide

COLLINS GEM
SEASHORE

COLLINS GEM
TREES

COLLINS GEM
Understanding **DREAMS**

COLLINS GEM
WILD flowers

COLLINS GEM
WINE Dictionary